AF762897

The international paper trade

The international paper trade

TOM BOLTON

WOODHEAD PUBLISHING LIMITED
Cambridge England

Published by Woodhead Publishing Ltd,
Abington Hall, Abington,
Cambridge CB1 6AH, England

First published 1998

British Library Cataloguing in Publication Data
A catalogue record for this book is available from the British Library

ISBN 1 85573 252 1

Designed by The ColourStudio (jacket).
Typeset by BookEns Ltd, Royston, Herts.

Contents

Preface

The *International Paper Trade* is a broad title for this book, but it does far more than discuss paper import and export. It discusses the whole spectrum of the pulp and paper industry. It is the intention of the book to give its readers an international overview of the paper trade. There is much factual information, presented as a comprehensive series of graphs and histograms. In fact, there are over a hundred figures to assist the comprehension of factors describing such a diverse industry. The book falls into three major sections: the fundamentals of the paper business; drivers of change and their effect; and the changing nature of the paper business.

Chapters 1–4 give a brief outline of the history and the technology of the industry, and show the production, consumption, import and export levels in the major regions of the world, over the past 15–20 years.

Chapters 5–8 examine the factors that are driving the industry today. They consider the forest resource, where the forests are to be found and what the principal forestry issues are, and they quantify the relationship between forest utilisation and paper consumption. The principal raw materials used in pulp and paper manufacture are also discussed. Environmental issues have a particular impact upon the paper industry, and waste recycling and the use of recycled fibre are discussed in some detail, as are the main environmental issues facing the industry. The importance of good public relations is highlighted.

Chapter 9 considers the changing paper business and what the next decade holds for the industry. What are the challenges that the trade faces, and what is the way ahead?

As an aid to the busy reader, each chapter begins with an outline of the subjects covered in the chapter and, where appropriate, finishes with a brief summary. Finally, Chapter 10 provides a concise summary of the main points dealt with in the book.

The book is generally designed to be of value and to appeal to the busy reader in the industry and its allied trades, who needs a good understanding of the paper trade. It is also aimed at those wanting a quick, in-depth study of the trade, because they are engaged in specific important, but short-term projects about the industry such as bankers, financiers, insurers and lawyers.

In compiling the book, I have drawn upon an extensive knowledge of the industry, acquired from working in a wide variety of senior appointments in different locations, both in the UK and overseas, over the past 40 years. I am indebted to many friends, colleagues and organisations for their help, comment and permission to use their material. I am particularly grateful to my colleague Clive Capps, who was the inspiration behind the concept, and who helped plan the scope of the publication. His speciality is raw materials and the environment, and I am indebted to him for his help and advice, and for the provision of some material.

Pulp and Paper International (PPI) have a publication *Facts and Figures* and I am grateful for their kind permission to use information from the *PPI International Fact and Price Book 1996* as a source of statistics from which many of the figures have been drawn. As a 'fact' book their publication is most valuable for anyone wishing to pursue further a requirement for more factual detail of production, consumption, import and export in the many countries of the world, of paper, pulp and recycled fibres, and of the principal players.

Another important source of information has been the monthly publication *Paper Technology*, the official journal of the British Paper Industry Technical Association which, together with their conference papers, have also provided a valuable source of information, and I am indebted to their Executive Director, John Clewley, for his permission to use the information.

Finally, every endeavour has been to make sure that the book is accurate, using information obtained from a variety of sources, and any comments that are made, and opinions passed, are done so in good faith. However, the book is influenced by the views of the author, and neither he nor the publisher can be held liable for any consequences of any action taken or for any commercial decisions based on or attributed to this book.

Tom Bolton

CHAPTER

1

Overview of the paper industry

The broad parameters of the international paper business are introduced and the scope of the industry is described in terms of tonnage and value. It consumes about 173 million tonnes of pulp, 100 million tonnes of recovered fibre and 276 million tonnes of paper world-wide, with a global value in 1995 of around US$540 billion, and probably as much again in supporting businesses. The history of papermaking is briefly dealt with, progressing from papyrus in 3000 BC, to hand-made sheets and then to the very big machines and outputs of the twentieth century. The link with sophisticated civilisation is described, followed by a description of the various paper classifications in the business. The chapter finishes by considering the importance of the international paper trade in comparison with other international trades.

The scope of the industry

The view of 'trade' taken in this book is not the purist view of macro import and export between regions. The book considers all trade, both within the country and between countries and regions. It is difficult to appreciate the full size and complexity of the international paper trade. This introduction provides an overview against which the succeeding chapters can be seen. Think 'paper' and you think of the main areas in which you come across the commodity. Newsprint and tissues are self-explanatory, except that tissues include toilet tissue, hygienic products, handkerchief tissues and other varieties. Corrugating papers are visible in supermarkets and off-licences, as the corrugated case workhorse, which carries bottles and tins, and they may also be considered in conjunction with packaging paper and packaging board. Printing and writing grades form the bulk of stationery, advertising materials and magazines, while speciality grades are familiar to us through cigarette tissue, or photographic papers or adhesive labels, or even as laminated working surfaces.

Apart from a small subsection dealing with converting, the subject of the conversion of paper to end-use products is considered to be outside the scope of this publication. Also outside the scope of this publication is the multi-billion dollar trade resulting from other uses of timber made available from the forests.

The simple outline of the component parts involved in the international paper trade is shown in Fig. 1.1. This is a simplistic chart showing the amount of materials consumed in the paper chain. Paper is made from cellulose fibres. They are contained in wood as cellulose and hemicellulose, with various largely unwanted materials, such as lignin and bark. The industry therefore embraces forest management and it uses 650 million cubic metres of wood from the forests to produce 173 million tonnes of pulp. It uses this, together with 100

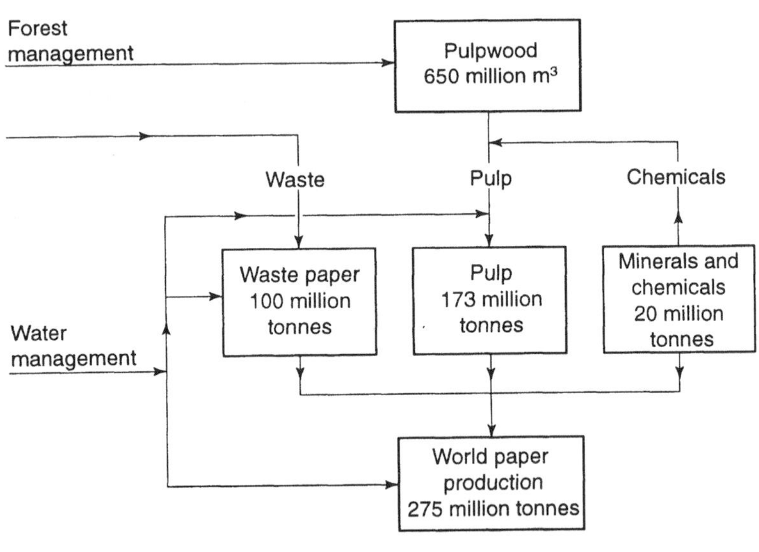

1.1 The international paper trade by tonnage in 1995. Figures are based on crude estimates. Note that it is inappropriate to quote volume figures for chemical utilisation.

million tonnes of waste paper, an all-important raw material of production, driven initially by economic pressures, but now driven by ecological and environmental considerations. Waste-paper management is now an essential part of the business, and the industry's usage exceeded 100 million tonnes for the first time in 1993.

The other principal component is minerals and chemicals. Mineral procurement must be considered, for paper is a principal consumer of the minerals kaolin (china clay), calcium carbonate, talc and titanium, all amounting to more than 20 million tonnes annually.

This describes broadly the segmented raw materials used in the manufacture of paper. Some 276 million tonnes of paper on the world scene is traded annually. This has grown at an average of around 3% per annum for the past 20 years, and has a crude value at, say, US$1200 per tonne of paper, of a trade of just under US$350 billion, giving a total trade value of paper and consumable raw materials used directly in its manufacture, and crudely estimated, of US$540 billion. This is laid out in Fig. 1.2.

In calculating these figures, the technologist could go into detail about the inescapable waste of moisture in the wood, in the pulp, and in the fibre entering the process, together with a small gain of a few per cent moisture content in the final paper, but although this precision is important for technical purposes, these refinements have not been taken into account here, since this section provides only a crude estimate, designed to show an order of magnitude.

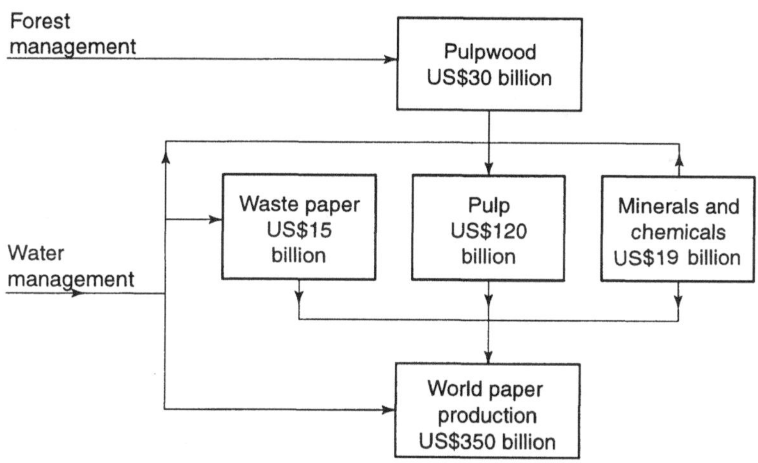

1.2 The international paper trade by value (US$) in 1995. Figures are based on crude estimates. The total direct paper trade value is US$ 534 billion (pulp wood, waste paper, pulp, chemicals and paper). The indirect value, e.g. machinery and clothing, is not included.

Besides the value of the raw materials and the paper, there is also the indirect value of the machinery supply industry. Paper is made using complex machinery, and a state of the art paper machine today is the size of a rolling mill in the steel industry. The paper machine may be 10 metres wide, produce paper continuously at approaching 2000 metres per minute for 24 hours of the day, seven days a week and, apart from the occasional shut-down, for 365 days per year. Such a machine will produce, say, 250 000 to 300 000 tonnes per annum of paper.

Such a production unit, including services and buildings, will cost, on a green-field site, approaching US$750 million. Then again, the cost of equipment for pulping fibre must be considered, and for virgin fibre, this could well add a further US$400 million to the cost of plant in an integrated mill which would bring the cost to in excess of one billion US dollars. If recycled fibre is used in place of virgin pulp, the cost would be appreciably less, say around US$100 million, leaving the total below the billion dollar level. Supporting this production is a variety of peripheral industries, some dedicated largely to pulp and paper, such as engineering and machine clothing companies, and some general industries such as the dyestuffs industry, or general chemicals or water management, where the use, reuse and clarification of water is again of vital importance. These international supporting businesses are again substantial. Estimates of their value are not included in Fig 1.2. Figure 1.1 has been redrawn as Fig. 1.2, where the statistics are now re-estimated as values in millions of US$. Once again it must be emphasised that these figures are just coarse estimates, setting the scene for the chapters that follow. The figures that are shown in Fig. 1.2 deal with the product and raw material value of the industry. Their purpose is simply to offer an indication of the value of the paper trade to the world's economy. They arrive at a crude estimate of US$534 billion, contributed by the trade in pulpwood, waste paper, chemicals and minerals, pulp and paper.

The history of paper

In 1995 the world demand for paper was calculated to be 276 million tonnes, and this is expected to rise by around 30–35% to near 350 to 375 million tonnes by the year 2005. Such a requirement needs big and sophisticated machinery for its production. This is a far cry from the way it all began. Although 5000 years is the total time span for these enormous changes, they did not get properly under way until around 200 years ago.

The journey of development took in stone tablets, clay, skins and other materials until we come to papyrus known in Egypt, somewhere around 3000 BC. It is this material that was found carrying hieroglyphics and providing a record of the Egyptian civilisation. Civilisation demands the written word, which requires a simple medium on which that word can be carried. The Chinese papermaker Ts'ai Lun made a pulp from fibres: he used material from which nets and ropes were made. He boiled them and strained them over a frame, and produced a sheet of paper which was then pressed under heavy weights. This was the beginnings of papermaking. The date was around AD 100. He went on to use other plant fibres and it is interesting that some of the materials he used are still used today, although wood fibre now predominates. The raw material that he used was fibrous, made up of cellulose and hemicellulose. His product was a hand-made paper.

Hand-made paper was the only way of producing paper until well into the seventeenth century, although improvements in techniques and in the chemistry of the process were introduced. Paper was sized (treated with a gelatinous solution) to take inks, bleached for whiteness, and coloured, but still the paper was hand made. The output from a paper mill was measured at just a few hundred tonnes annually.

Then came mechanisation. In 1799 Louis Robert conceived the idea of a paper machine, but like most good inventions it was not immediately successful – the best ideas rarely are – and it was further developed by Bryan Donkin, and the brothers Henry and Sealy Fourdrinier. The machine became known as the Fourdrinier machine, and it still carries this name today. It was a huge step forward. The output from the first machine was 600 lb (272 kg) in 24 hours. But how much further has the industry travelled! Today, a newly installed, up-to-date machine will produce at a rate of around 3000 times this.

The processes remain fundamentally the same. Trees are debarked mechanically, and although some low-quality pulps are not cooked, so they retain their lignins, which adversely affects their ageing properties, over half of the cellulose fibre used is cooked to remove impurities such as lignins and resins. (Lignin is a complex polymer in the cell wells which make a plant rigid and woody.) They are often, but not always, bleached, and the pulp so formed is the basic raw material for paper production. The pulp is then mixed with vast quantities of water. There are three reasons for this. The first is so that strength bonds develop between the hemicellulose in the fibres, and on this depends the final strength and thickness of the paper. The second reason is so that the slurry can be readily moved around by pumping. Thirdly, having got the suspension of fibre and other additives to the paper machine, the water is used to spread it out uniformly - very uniformly - after which the process becomes the removal of water. Every tonne of paper manufactured uses around 100–500 m^3 or 22 000–110 000 gallons of water, although of course more than 90% of the water will then be reused, so reducing the freshwater load by 90%.

The year 1800 saw the first paper machine. The design was improved, the machines made bigger, and by 1950, just 150 years later, with the industry at last fast-changing from a craft-based to a technologically based industry, the machine was say, five times as wide, and capable of running at over 500 feet per minute (150 m/min). The 24 hour output had risen to 50 tonnes.

In the period from 1950 to 1980 much rationalisation and reorganisation of the industry took place. Paper was fast becoming a commodity item, losing its specialised nature, and it was under great price pressure in a situation where there was more often than not spare manufacturing capacity. After all, a new paper machine produced a considerable amount of new production capacity, and although the business plan paid regard to competitors it was not unusual for two companies in the same market to invest simultaneously in new output to service much the same market. This 30 year period showed more sophistication and invention which was manifest in greater outputs, and there were more manufacturers trying to use old machinery to make fine papers, as opposed to mass-produced commodity papers. Consequently there was a great deal of rationalisation and mill closure, while production costs were driven down, and more and more of the process was automated.

A further innovation was the widespread and growing use of waste paper replacing wood pulp. It was originally introduced for economic reason, but the environmental demands to replace wood pulp with a recycled material have led to the production of a high-quality waste pulp, which can cost as much as the wood pulp it replaces. The substitution of wood pulp by waste paper has risen globally until it was around 38% in 1995.

Where are we today? The paper machine is as big as a steel rolling mill. It is 10 metres wide, 800 metres long and capable of producing paper at up to 2000 metres per minute. It is computer - and operator - controlled, and it is highly sophisticated. Its output is 250 000 tonnes a year.

The role of paper in society and business

Paper consumption is always considered to be an indication of the sophistication and literacy of a nation. Consider Table 1.1 which shows the statistics of paper consumption per capita in the major regions of the world today.

Table 1.1 Per capita consumption (kg per capita)

	1973	1983	1994
Europe	72	95	97
N. America	234	302	322
Asia	9	20	24
Australasia	98	132	136
Latin America	17	26	29
Africa	3	5	5
Rest of World	32	45	48

North America heads the list, and one has only to visit the USA to start to appreciate just how widely paper is used. It is not just the number of ways in which paper is used, it is the sheer quantity. The Americans lead the way with huge multi-sectioned Sunday papers, although this says nothing about the quality of the publications. In America the employees' remuneration package is generally high, and so the consumables bought are high, and this in turn flows on to packaging, boxing and advertising. It is no great surprise that printing and writing grades are one of the biggest growth areas, since more goods sold means that more suppliers wish to put their goods in front of their market, and so more advertising material is supplied. It is particularly interesting to look at per capita consumption figures, indexed by country to the 1984 per capita value, and shown in Fig. 1.3. They show Asia and Latin America leading the per capita growth in consumption, at a level well above world growth figures. These figures show some similarity to the figures for population growth taken over overlapping periods of time (Fig. 1.4).

One may imagine the figures of per capita paper consumption for a high-population country, without, sometimes by choice, Western sophistication, for example China, India or South America. The per capita consumption will be increasing at a greater rate than in a high per capita consumption country. In fact some believe that by the year 2005, China will have become the second largest consumer in the world, after the USA. The projected population of China is 1500 million by 2025, a three times increase since 1950.

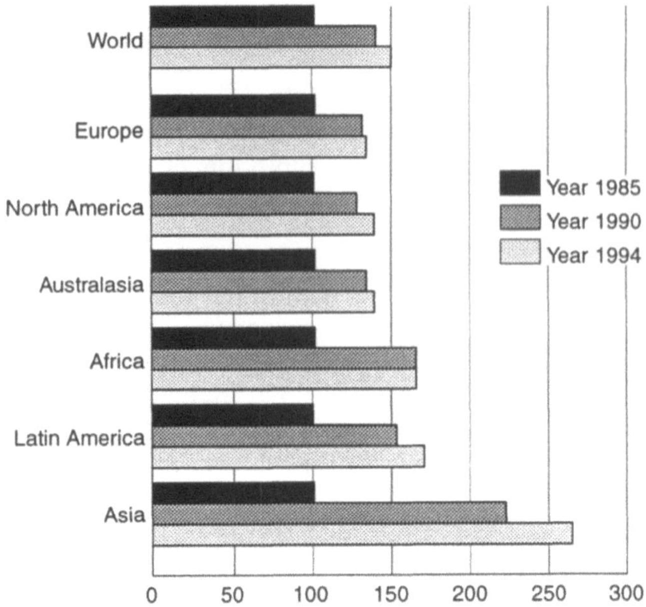

1.3 Per capita consumption by country, indexed to 100 in each country in 1985.

It can be seen that the availability and the use of paper is some measure of developing affluence: perhaps a better word is sophistication. It is ironic that at the same time, paper has from time to time been blamed for national disasters. It was considered in some quarters that it played a large part in carrying the great plague of London. Rags are made from cellulose fibres and provide an excellent papermaking raw material. It is suggested that rags from people with or in contact with the plague were used to make paper and so carried the plague. It is a hypothesis that may be treated with some scepticism in this early example of recycling.

Modern packaging papers are not only used to provide eye-catching wrappers. They are important in preserving food in hygienic conditions for extended periods of time. Were they not used, the food would deteriorate and be wasted before it could be consumed.

It is interesting also that modern technology has had a positive effect on the demand for paper. In the 1960s erudite discussions among the technical and marketing community suggested that the advent of the computer would bring with it the paperless office. Look at the way in which paper consumption in the office has increased over the past 10–20 years: spread sheets, computer papers copying papers, letterheads, etc. Computers make the use of paper easy and

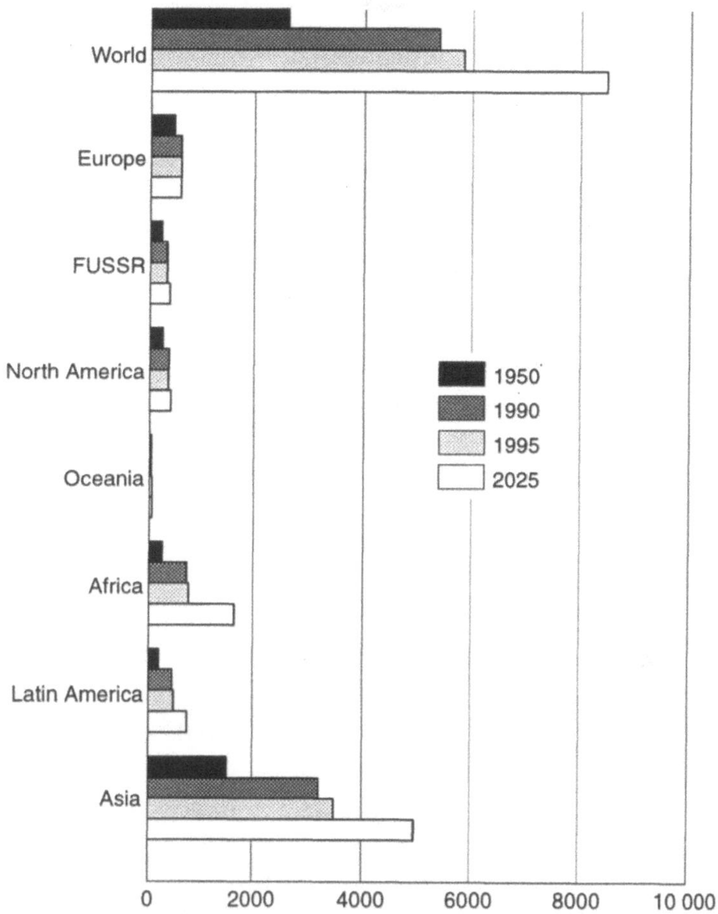

1.4 World population (millions).

also desirable, hence they lead to greater consumption. It is interesting to speculate on whether or not this will continue.

In the past few years widespread use of e-mail has been the latest development to come on stream and be commercialised. But people still like hard copies, and it is difficult to see this development bringing any overall reduction in the paper trade.

Grades of paper

So far the discussion has been in general terms about paper, but what has not been considered is the widespread use and classification of paper. Broadly, paper is classified in seven groups:

1. *Newsprint*: used for mass media contact through newspapers and for some very inexpensive advertising.
2. *Corrugating*: used as the inside fluted layer of corrugated boxes and packing materials.
3. *Packaging papers*: papers of quite high quality used for packing and wrapping. They will have functional qualities in protecting food besides the ability to be printed to a high quality, for example, for cosmetic advertising. In fact some of them may be coated to give them a high degree of moisture impermeability or grease-resistant properties or to accept high-quality printing.
4. *Packaging boards*: display cards, cosmetic and cigarette boxes all of very high quality, down to thick functional boards.
5. *Printing and writing grades*: book papers, writing papers, continuous stationery, carbonless copying papers. Light-weight coating papers used widely as magazine and advertising papers.
6. *Tissue grades*: papers for toilet tissue and sanitary grades, handkerchiefs.
7. *Other grades*: speciality grades for niche markets. For example, photographic papers, cigarette tissues, laminating papers, artist's papers, adhesive labels and so on.

These classifications are universally accepted and they will be returned to in a later chapter when world consumption by grade is considered.

The importance of the paper trade

The paper trade is not often thought of as a major international industry, yet that is exactly what it is. It is not easy to find meaningful, comparable figures for the value of a world industry to the world economy, but the following figures have been researched to compare paper briefly with other industries that come readily to mind. First of all, one can assemble a 'world fibre league table' (Table 1.2) which gives an order of magnitude value of fibre volume and value in around 1992. This shows clearly the important and dominant position taken up by wood pulp fibre in the world fibre league table.

In arriving at a rough value for the pulp and paper and other industries, it must be emphasised that the figures can once again only indicate an order of magnitude. Nevertheless a figure of US$500

Table 1.2 World fibre league table

Fibres	Million tonnes per annum	US$ billion
Wood pulp	165	99
Cotton/wool	17	41
Synthetics	15	35
Non-wood	5	2
Asbestos	5	2½
Glass	2	6
Mineral	2	?
Carbon	30	¼

Table 1.3 Crude listing of commodity output in 1992

Commodity	Tonnage (million tonnes)	Unit value US$/tonne	1992 value (billion US$)	Estimated value US$ billion
Crude oil	3000	137	411	520
Pulp and paper	412	600–1200	395	435
Steel	735	120	350	440
Tobacco	8	4000	32	40
Iron ore	950	32	31	40
Aluminium	19	1050	20	25
Bauxite	110	145	16	20
Wool	3	2440	7	9
Rubber	6	930	6	8
Cocoa beans	2	1250	3	4

Source: based on UNCTAD (1995).

billion has been estimated, based on a paper-selling price of US$1200 per tonne, and a pulp-selling price of US$600 per tonne. Table 1.3 offers a crude order of tonnage and value of some main commodities, based on 1992 figures derived from statistics produced by the *United Nations Conference on Trade and Development Commodity Yearbook 1995*.

Although it has been emphasised that the approach taken in deriving this table has been a very crude one, nevertheless the order of the commodities and the order of magnitude of production amounts are in the right order. It is when one comes to address value that the influence of world economics makes the results a little less certain.

Notwithstanding this, it is of importance to note that the paper trade ranks second with steel in the list, and that the pulp and paper value estimate together is close to the value of the crude oil industry in sales value per annum. Leaf tobacco is the next contender in the rankings and that only approached 10% by estimated value of that of the paper industry.

This of course takes no account of the social significance of the industry. To quote the International Institute for Environment and Development (IIED): 'Paper products are used to communicate ideas, to inform and instruct, and to record events and cultural information. They also play a key role in human health, in the form of food packaging, medical dressings and sanitary products.' A strong point in favour of the industry is its heavy reliance on renewable resources, and therefore its management of them. The paper trade is significant as a world industrial producer and trader. All in all, paper has a strong, meaningful and sustainable position as a major world industry.

Summary

The international paper trade is a multibillion dollar industry producing 276 million tonnes of paper a year, at a total business value in excess of US$500 billion. In over 5000 years it has changed from a small craft industry to one of sophisticated big machinery, with modern machines each producing around 250 000 tonnes per annum. It is an industry upon which many others are highly dependent. Its ramifications are wider than just the pulp and paper businesses and so far, paradoxically, moves to the 'paperless office' have resulted in increased quantities of paper being used. Whether or not this continues remains to be seen.

Preservation of the forest is a major concern today. A later chapter will consider the influence of the paper industry on the forest, where wood provides a vital raw material for the manufacture of paper. This in its turn makes the industry consider the continuing provision of healthy forests, which will in the last analysis ensure that the paper industry makes a real contribution to the economic and ecological development of the world's forests.

The paper trade uses raw materials that can be regenerated, and the product can largely be recycled. It is a major international industry.

CHAPTER

2

Manufacturing and conversion

The processes in paper manufacturing are outlined in this chapter. Manufacture starts in the forest from which the majority of raw material is drawn, and progresses through the various stages from the manufacture of pulp, to the paper machine and then through the finishing operations to the consumer product. The chapter refers to the future development of the process, and makes comment on the profitability of paper manufacture. Materials other than cellulose-based raw materials are briefly described, and there is a discussion of some of the conversion opportunities in a number of industries built around paper and to which the industry is one of the main suppliers.

Introduction

The technology of pulp and paper manufacture is quite a complex study. This chapter is designed to give those not normally concerned with the technology an outline appreciation of the manufacture of pulp and paper. It introduces the principal processes and the main raw materials used, and concludes with a commentary on some converting industries reliant upon the product.

Paper is usually manufactured from wood fibres, derived from virgin wood pulp or from recycled fibres, in a continuous process.

Trees from the forest have a covering of bark, which must first be removed and the wood chipped before the cooking process which removes lignin and resins from the fibres and helps to separate them. The amount of lignin so removed is 21–29%. What remains, and is the principal component of paper, is essentially cellulose, a linear polymer of carbon, hydrogen and oxygen, and hemicellulose, which has a similar chemical structure, but which has the property of promoting bonding of fibres when the hemicellulose fibres are in close contact in a water suspension. The principal components of a wood fibre are:

- Cellulose 44% (43–45%).
- Hemicellulose 31% (28–34%).
- Lignin 25% (21–29%).

As the speed of manufacture has increased, so has the degree of sophistication of the processes. As we have seen, a single completely integrated production line on a green-field site, processing wood and producing paper could cost in the order of US$750 million.

Put quite simply, the process takes forest wood, frees it from its bark and lignin impurities, and separates the individual fibres. The fibre is treated in a number of ways, using much energy, to enable it to develop its required properties of bulk, strength, uniformity and colour, using water both as a carrier and to allow strength properties to develop. The process then becomes one of uniformly spreading out the fibre suspension from the pipe through which it is pumped, to the width of the paper machine, and then removing water as inexpensively as possible, by draining, squeezing and steam drying, until the final sheet of paper is produced. The continuously made paper is then cut into reels or sheets for subsequent use and conversion. This flow line is shown diagrammatically in Fig. 2.1 where the main process line is shown in bold type with some principal additives being shown in italics. It is appropriate to consider these processes in a little more detail.

Forest to pulp

The first process constitutes an industry of its own. It starts in the forest which is ideally well managed and sustainable, and where trees are felled and taken to the pulp mill, in many cases via the timber yard where the best timber is taken for use in furniture-making and building. In the pulp mill the bark is removed from the wood which is

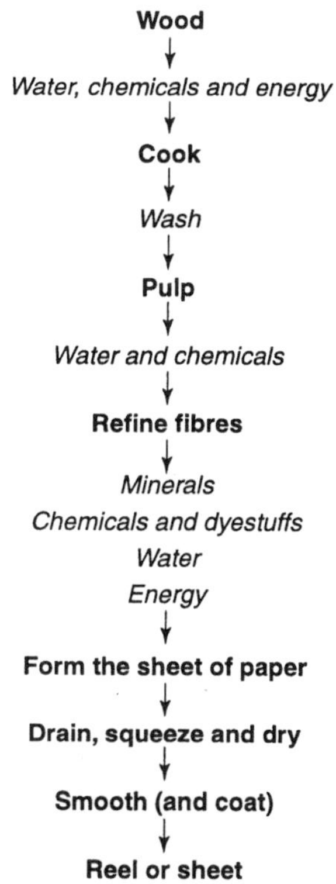

2.1 Wood to paper: a flow diagram.

then chipped, consuming energy. Depending on the process, wood could be mechanically treated to separate the fibres but leaving in the wood impurities such as lignin, or chemically cooked to remove these impurities and to separate the fibres. The cycle times in the cooking operation range from continuous processing to batch processing for 8–36 hours cooking per batch. Typically, continuous production is becoming the mode, but 8–12 hour batch processing is not unusual.

Traditionally pulp mills are found near to a forest and a water supply, the latter often providing relatively cheap energy in the form of hydroelectric power, although with the 'urban forest' giving a source of recycled fibre the position is now changing and some new developments are located within a reasonable distance of the city from which the recycled paper is drawn. The paper mill is frequently integrated with the pulping process, but there is a considerable part of the industry that uses pulp from a separate site far distant from the paper mill, which may use bought-in market pulp. Large volumes of water are used in the pulping process which more often than not also consumes chemicals for cooking and bleaching, and steam for cooking. Once the cooking is complete, the pulp is washed to remove residual chemicals. Some of the pulp may be pumped to storage tanks for subsequent processing. Another part of the pulp may be taken through a bleaching process and the white pulp washed, and stored for immediate use, or dried, lapped and baled, and stored or sold as market pulp.

It must be emphasised that from the early days of making pulp from wood in the Northern Hemisphere, such an important industry has, in the main, made secure its future wood supply, and so the forests have by and large been well managed and regenerated, although there are some areas, for example in Russia, where this is not always the case. A similar tradition is also developing in some parts of the Southern Hemisphere. This important environmental position needs stating given the current political atmosphere.

Pulp to paper: the paper mill

Pulp refining

The next series of integrated processes, shown schematically in Fig. 2.2 on page 21, which takes place at the paper mill, must be considered. The fibre is again slurried (unless the mill is in line with the pulp mill in which case it will be supplied as a slurry), and stored in huge containers, known as chests, from which it is pumped to a refiner. It is then refined by passing it between the steel-toothed plates of fast-rotating machines. These refiner plates are set only micrometres apart, and their purpose is to ensure that every fibre in the slurry is treated uniformly to develop its required properties, mainly strength, bulk and porosity. This part of the process also consumes a substantial amount of energy.

When the treatment is complete, the fibrous slurry is further diluted with more water and chemicals added – for example, clay, size, dyestuffs – and the slurry is passed on again to the paper machine.

Water plays an important part in this stage of the process. To give some idea of the amount of water involved at the refining stage, there may be 10 parts of fibre (typically 4 to 12 parts) to every 100 parts of water. At the final dilution point, before the sheet of paper is formed, there are around 5 to 10 parts of fibre to 1000 parts of water. Put into perspective, for a modern paper machine producing 250 000 tonnes of paper a year, a water usage of around 11 000–55 000 gallons per minute, or 50–250 m^3 per minute, is normal. However, the freshwater requirement is usually reduced by at least 90% by its reuse around the process system.

Sheet formation and drainage

The water will help to develop the strength and bulk of the fibres through enhancing their ability to bond. It has transported the fibres around the mill by means of pumps and pipes, and after the diluted slurry has been cleaned with centrifugal cleaners and screens, it now plays a fundamental role in spreading the fibres out to a high degree of uniformity. This part of the process is of great importance and sophisticated equipment is used to distribute the fibre and water carefully and uniformly across the width of the paper machine. The equipment is called a flow box and may cost up to, say, US$1 million. It spreads the fibres out from a one metre diameter pipe to a uniform web of paper some eight to ten metres wide. This good uniformity is of crucial importance, since it assists the way in which the paper is transported through the paper machine continuously and without breaking, and it contributes to the paper's efficient performance in subsequent processes and conversions.

The water has now served its purpose and must be removed so that a sheet of 'dry' paper can be made, leaving in the paper just sufficient water to ensure the paper lies flat and is in equilibrium with the humidity conditions it encounters in subsequent processing. Water removal is then carried out in a series of processes of increasing expense. The first is by drainage of water through a very fine mesh synthetic fabric, seamed into an endless belt. After this, water is removed by dynamic drainage systems, or hydrofoils, and then by suction through a series of vacuum suction boxes. At this stage the water to fibre ratio has become something like 1:1, or 50% moisture content. Much water has been removed, and will be immediately reused in the process, together with any fibre, loading and heat that it may contain. Nevertheless, there is still more water to be removed.

Pressing and drying

The paper is automatically fed on through the machine to a series of presses that squeeze out the water, which is taken away with the help of absorbent felts, so that the water is not reabsorbed into the paper. At this stage, the fibre to water ratio is of the order of 1: 0.75, or 40–45% moisture content.

Then comes the last part of the water removal process, where the paper is dried on steam-heated drying cylinders or drums to which it is held in close contact by a series of drier screens, which are continuous, open meshed, woven synthetic fabrics. The process is assisted by moisture vapour removal systems designed to promote evaporation among the drying cylinders, and to prevent pockets of saturated air from inhibiting the evaporation process, and causing the product to lose its uniformity.

The paper may then be rewetted at a size press, to improve its cross web moisture uniformity, and its printability by the application of starch and water, before further drying and reeling at a fibre to water ratio of 12:1, about 7.5% moisture content.

Smoothing and coating

Smoothing may take place either on the machine or off the machine in a separate process, or both on and off the machine. It is carried out on the machine, by a high-intensity soft nip calender, where the paper is fed between two rolls, heavily loaded on their ends. The bottom roll has a camber adjustable using a floating roll system so that the nip between the rolls is parallel under different conditions of loading. It is imperative that the nip pressure is always uniform across the roll. This is so that any smoothing application does not detract from the uniformity of the cross web thickness profile of the paper, and it is for this reason that expensive floating rolls are used.

Off-machine smoothing is carried out on a supercalender. The supercalender is a sophisticated piece of equipment that comprises a stack of rolls, or bowls, of different materials and hardness. They are run at high temperatures, and the paper passes through them at speed. The process confers properties of smoothness and gloss, and causes a reduction in paper thickness. Again floating rolls may be used in the calendar design in order to adjust nip pressure and make it uniform across the face of the roll.

Coating may also be applied on or off-machine as a surface layer, where typically coat weights can increase the total mineral content to, say, 30% in the case of an art paper or as much as 40% in the case of a light-weight coating grade. Off-machine coating is more versatile, but on-machine coating allows a good coating application for a commodity paper at a lower cost. There are a number of coating application methods, but all involve applying a layer of coating mixture, usually containing either china clay or calcium carbonate with an adhesive of starch or latex, and with other chemicals to control the application of the coating mix. Inevitably water is added with the mixture, and so a drying system needs to be built into the machine. Usually a cylinder - air cap system, possibly supplemented by infrared heating, and perhaps with some air float dryers, is used, although again it is more common to see floater dryers used with off-machine coaters.

The on-machine coating section is again a large machine integrated into the paper machine, and it needs all of the sophistication of the paper machine, requiring its own sectional drive and process and computer control to allow high performance. In some plants, the coater is off-machine, and in these cases speed and size need careful study before the machine is purchased, since the coater will have to be able to produce at the same rate as the paper machine.

Different parts of the converting industry apply a number of specialised coatings to different base papers for different applications - for example silicon for release papers, or polythene for photographic papers. It is in manufacturing these speciality papers that certain companies and mills have developed specific expertise.

Reeling, winding and slitting

Each machine has a reel-up mechanism and a winder. The latter is used to take out the paper marked to be defective, to trim to the finished width and to prepare good defect-free reels for subsequent conversion. The winder may be followed by a slitter which will then slit reels to customer width and length, usually discharging many webs in sets. These sets may then be automatically wrapped, and used for subsequent conversion, usually outside the paper mill.

Sheeting

The sheeter takes multiples of reels from the winder. There can be as many as eight or ten at a time. They are all cut together, multi-ply, into the sheet size required by the customer. The sheets are stacked, possibly guillotine trimmed and automatically counted and wrapped.

Some specialised sheeters are available to sheet into cut sizes, for example A4, and to count, pack and wrap in one operation.

The machine drive

The paper machine comprises a number of sections which are all driven by a sophisticated electronically controlled drive.

During the water removal processes, the fibre mat shrinks all the time, while at the same time stretching a little under the tension of the fast-moving web of paper as it is transported through the machine, and care must be taken to drive the interlinked sections of the machinery in such a way that these dimensional changes are allowed for. The sectional electric/electronic drive of the machine is of the greatest importance to the smooth, trouble-free running of the paper machine. It must be able to drive the sections of the machine absolutely uniformly and at constant speed, at all speeds from crawl, when the machine is fed with paper, to 2000 metres per minute in a well-controlled, linked manner, with continual control of tension down the machine. The drive is at the heart of the continuous process, and the machine must run steadily at a precisely uniform speed, under all conditions of acceleration and deceleration, and without causing paper breaks.

Process and computer control

Just as the drive will nowadays use sophisticated computerised control systems, so there are numerous subprocesses in the machine system that are similarly controlled. There are, for example, level controllers and magnetic flow meters at numerous points in the system, controlling chest level, flow, pumping rate of additives, fibre to water ratio, mineral addition rates, retention aid supply and very many others. These parameters, together with measurement and control of process parameters of weight, thickness, moisture content and mineral content, are linked by computer control, both in small loops and in sophisticated integrated systems. They have sensors to measure desirable paper properties both across the machine width of paper and along its length. The system is supervised by the operatives who are continually on the look out for the warnings and other information displayed from time to time by the various systems, so that they can take appropriate action.

A machine using all the available modern technology will be around 10 metres wide, and it will make paper at speeds of up to 2000 metres per minute. A paper that is 60 grams per square metre in weight will be produced at a rate of 30–40 tonnes per hour, or around 250 000 tonnes per annum. Figure 2.2 provides a schematic diagram of the process flow line.

The pulping and papermaking processes have been outlined and it is now appropriate to consider one or two of the subindustries that have grown that specifically supply the pulp and papermaking industry.

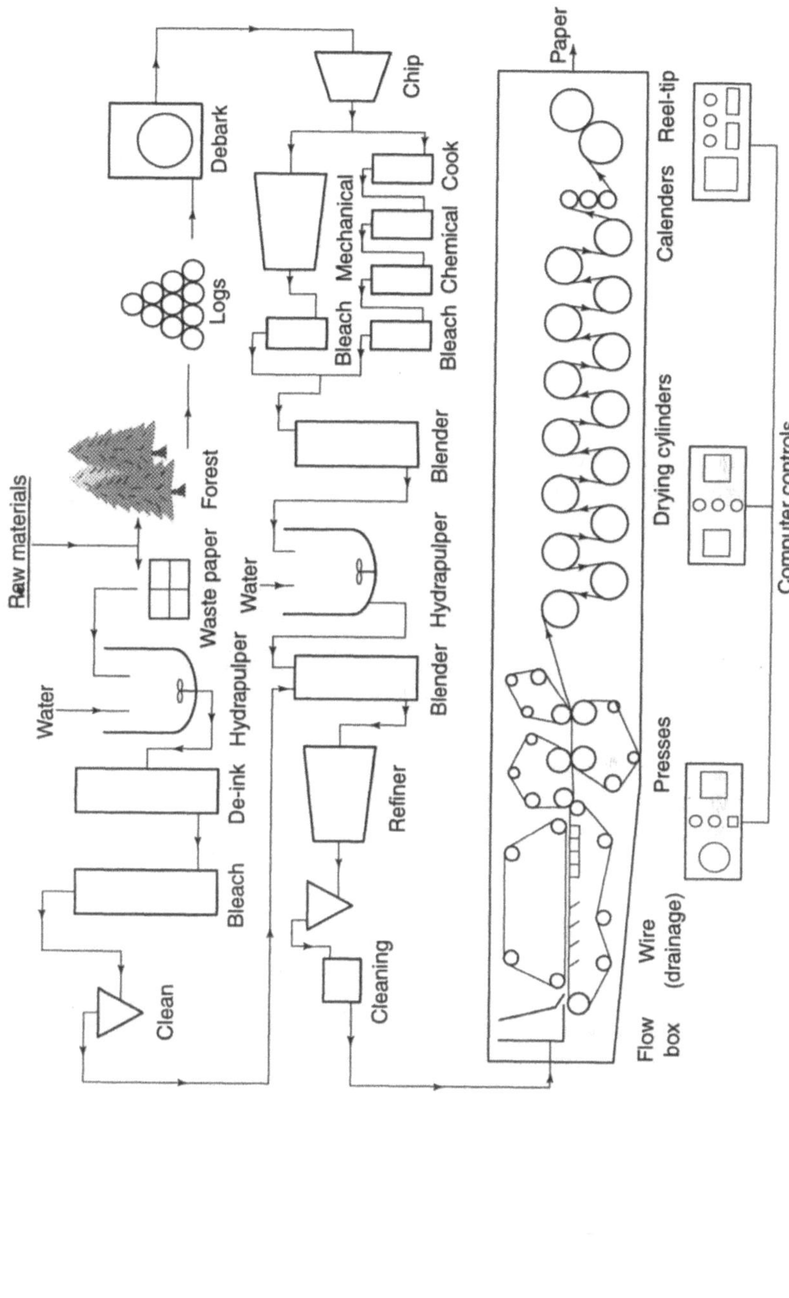

2.2 Schematic diagram of the process flow line.

Machine clothing

Passing references have been made to mesh fabric drainage systems, to absorbent felts and to dryer fabrics and screens. All paper machines have these items of equipment, which are known as machine clothing. They are disposable, very uniform fabrics and screens, each with a life that will vary from weeks to more than a year, depending on the application position, and local conditions. Examples in the next subsections are based on a lightweight coated paper machine 9 m wide and producing at 1200 m/min.

One complete set of clothing will cost around US$600 000–800 000, and it is usual to have at least two sets to back up the ones in use, with probably additional supplies of forming fabrics and press felts. It would not be unusual to have a machine clothing inventory of at least US$2.5 million.

Mesh fabric drainage systems

These are continuously woven synthetic fabrics, of different filament design and weave. They are made to a high standard of uniformity, and are designed to run in a continuous loop around the drainage section of the paper machine known as the machine table. The fabric must drain water from the web of paper absolutely uniformly, and without leaving a residual mark, or impression on the paper. The machine will have a top and a bottom fabric. The bottom position will consume four to six fabrics per annum and the top some three to four. The fabric life is important, not only because it is a costly consumable item, in its own right, but because to change the fabric will occupy some hours of non-productive time for the machine.

Press felts – absorbent felts

Presses are clothed with absorbent felts whose job is to absorb water squeezed from the paper and take it away. The water will be removed from the felt before the felt again comes into contact with the paper. These 'wet felts' are woven continuously or are seamed at the clothing manufacturer's plant. Again they are made to a high degree of uniformity, and changing them again means a loss of production time by the machine, probably of one or two hours. They are treated or 'conditioned' as they work, ensuring a continuing ability to remove water from the paper consistently and in a uniform fashion. The machines will each have a pick-up press and two other presses. They will be clothed with four press felts in all. These felts will be a pick-up felt and a bottom press felt for the pick-up press, and a single press felt for each of the remaining presses. Each position will use between 6 and 16 felts per annum, so that, typically, the presses will consume 35 to 42 felts per annum.

Dryer fabrics and screens

The principal task of dryer fabrics and screens is to keep the paper in close contact with the drying cylinders, so promoting efficient evaporation from the steam-heated cylinders, while at the same time being of an open mesh construction to allow steam removed from the paper to pass through them. They are put on the paper machine and they are seamed *in situ* using a fastener. They must not mark the paper with which they are in contact, and they must allow steam and water vapour to pass easily through them.

A typical drying arrangement of a paper machine comprises groups of each six or eight drying cylinders each arranged in four 'sections'. Each section is clothed with a top and bottom dryer screen. Here the clothing lasts for just under a year, so that around four sets of clothing are used every three years.

A light-weight coating machine will also incorporate an on- or off-machine coater, and this too will have two dryer fabrics.

The screens are run in conjunction with ventilation systems assisting the removal of vapour. The systems may be directed at the pockets of steam between the clusters of drying cylinders, or at the air above the driers. If a paper machine makes 250 000 tonnes of paper a year, and the moisture content is 50% on entering the driers, the drying system will handle around 250 000 tonnes of hot water released as steam from the paper.

A summary of this information is contained in Table 2.1. It is not surprising therefore that there is a major supply industry for machine clothing which has been built up around the provision of these expendable items. Finally, it is worth observing that each clothing failure causes unproductive time on the machine, and so it is usual to plan ahead for this time so that it can be combined with other maintenance work of the machine in a planned shut down.

Table 2.1 Paper machine clothing

	Position	No. of fabrics	Average life (months)
Wire part	Top	One	3–4
	Bottom	One	2–3
Presses	Pick-up	One	1–1.5
	Second	One	2
	Third	One	1–1.5
	Fourth	One	0.75
Dryers–machine	Four sections	Eight	9
Dryers–coater	One section	Two	18

Engineering

It has already been seen that the paper industry is capital-intensive, through high-cost pulping machinery, high-cost paper machines and supporting equipment, all of which contribute to the high machinery and therefore depreciation cost. It will also have been observed that, in reality, a paper machine is a series of quite complex, interlinked and sophisticated processes. As technology improves so sections and parts of the existing machine need to be replaced, maintained or updated. All paper machines have a continual investment programme, and some machines installed in the early 1930s are still operating, but are not recognisable today in their original form, since they have been operated under a constant programme of modernisation. But while speed can be increased, the original width of the machine usually cannot.

The number of engineering firms reliant on the paper industry for their livelihood is again large and many operate in specific specialist areas of the process, for example, vacuum box manufacture, ventilation systems and steam siphon pipes.

Paper machinery development

The development of the machinery to manufacture paper is continual, and innovative. Because there is the potential for great financial risk to a paper mill from introducing plant that has not been proved, dramatic development has been slow, but the past 15 to 20 years has seen, for example, the change from single wire forming to twin wire

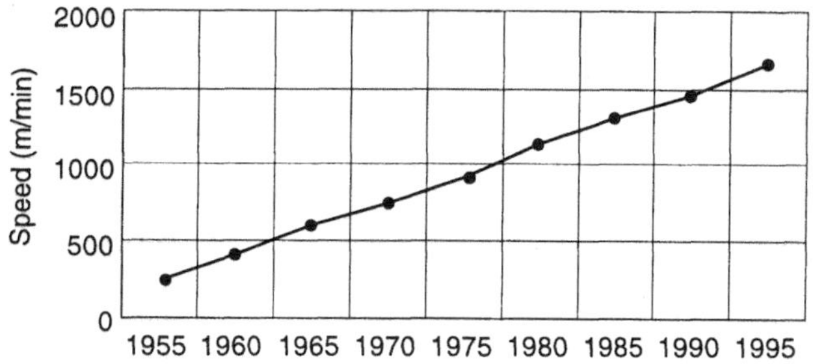

2.3 Paper machine design speed (m/min) 1955–1995. (Source: Valmet Machinery Manufacturers.)

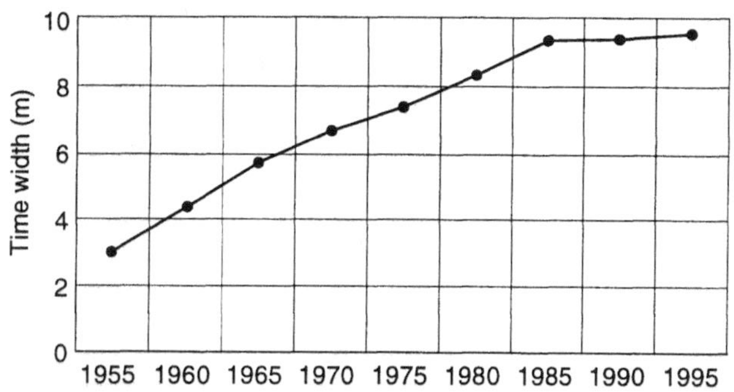

2.4 New machine trim width (metres) 1955–1995. (Source: Valmet Machinery Manufacturers.)

forming, and in some cases the move from two tier drying to single tier drying. These changes affect the traditional geometry of the machine considerably.

Nevertheless, the developments over this century have been quite spectacular. The paper machine has demonstrated six- or seven-fold increases in speed: Fig. 2.3 shows the speed trend since 1955. Machine wire width, and so the paper trim width, have also increased (Fig. 2.4) three or four times from 2–3 m to around 10 m since 1995. Both of these graphs have been derived from information from one of the machinery manufacturers, Valmet, about their machines. As may be expected, the raw data show a very wide range of figures and so these graphs show only trend lines.

Progress in the future is a matter of putting engineering development into all the other subprocesses on the paper machine. Certainly the increase of width seems to be reaching an optimum, but machine speed is still increasing. Barber (1992), in a paper entitled 'Latest developments affecting machine design', sees for example a newsprint machine of tomorrow being 11 m trim width, with a design speed of 2400 m/min producing 450 000 tonnes per year.

Profitability

The manufacturing industry is profitable, but the profits are cyclic in nature. The capital cost of modern plant is extremely high, and profitability often depends upon how the amortisation of the plant is dealt with and the way in which the loans or capital necessary to build it are handled.

The cost of manufacturing the product embraces both fixed and variable costs. The fixed costs of an operation are the people costs and the financial charges of one sort or another, for example, depreciation and taxes. The variable costs are, for example, operating costs - raw materials and chemical costs - energy and consumables, such as machine clothing.

Quite clearly, with such high-cost machinery, the machines and mills must run for as many hours as possible for 365 days a year, and seven day, five shift working is typical. Much effort is put into ensuring that the machines run without paper break and at their optimum speed. The paper machine closes only infrequently, and when it does become necessary to close it down, the downtime so generated is where possible fully planned and utilised for maintenance work.

Because of the method of shift working, only a small portion of a mill's employees may be seen in the mill at any one time. Typically an up-to-date, non-integrated, one machine paper mill of this size will have a total work-force of, say, 500 to 600 running the machinery, testing and troubleshooting, engineering, carrying out the essential administrative duties and resting off-site while the staff await their shift rotation. As far as the paper machine itself is concerned the actual number of operatives running the machine, but not the peripheral plant, at any one time will be around five, together with a shift supervisor.

One can also see that to run two machines of this size on one site would have great economies of scale, and this may account to some extent for the market being oversupplied, particularly when one considers the enormous step changes of supply that the introduction of a new production line engenders. This oversupply is detrimental to maintaining a good selling price, and this in its turn also has an adverse effect upon profitability.

A total of 24 000 people in the UK are employed in an industry producing 6 million tonnes of paper per annum. Using this statistic, and taking into account some overmanned operations in the East and the developing nations, one could project the total number employed directly in manufacturing to around 1 million. Extending this to include pulp production and the selling operation, could easily take that figure to 1.5 to 2 million.

Materials other than fibres

Water

As we have already seen, water is used extensively in the papermaking process. Mixed with water, the slurry of fibre and loading can be conveniently pumped around the mill, and a very dilute slurry is critically important in spreading the fibre out uniformly on the forming fabric. We have also noted the importance of water in assisting bonding between fibres.

The statistics pertaining to water consumption are these:

- One tonne of paper requires up to 10–50 tonnes of water after allowing for recycling 90% of the water within the process.
- 250 000 tonnes paper per annum is equivalent to 0.5 tonnes paper per minute.
- A modern paper machine (250 000 tonnes per annum) uses water at a rate of 11 000–55 000 gallons per minute or 50–250 m^3/min.
- If 90% process water is reused, the inflow of make up water is 1100–5500 gallons per minute or 5–25 m^3/min.

If the water is extracted from the river, it needs to be treated to a high level of purity through a water treatment plant. Impurities are flocculated (that is, formed into a bulky precipitate), settled and filtered before the water can be used in the process.

Within the plant, filters and save-alls are used both to recover fibre and to repurify water for immediate reuse. Then, before effluent is finally disposed of, it must again be purified, by having the maximum water extracted before the solids are disposed of, and the purified water returned to river, sea or sewer. Solid disposal is a problem, particularly in some countries where landfill sites are at a premium, and alternative disposal, such as incineration, is becoming important. The subject of disposal will be dealt with in a later chapter.

It can be seen that water and effluent treatment for the paper industry provides an essential sub-industry, requiring the supply of equipment and chemicals. This large sub-industry adds to the value of the paper industry and strengthens the economy of the country in which it operates, particularly if, as is almost certain, that economy already possesses suppliers to a water industry.

Minerals – fillers and coating

The world consumption of 20 million tonnes of minerals in the paper industry is worth around of US$4 billion. The slurry from which paper is made is varied to suit the paper grade being produced. The inclusion of minerals in paper improves smoothness and printability, makes the product opaque, and reduces its cost by replacing fibre. Also, minerals applied to the surface of paper by a coating process enhance the receptivity of the paper to printing inks, giving improved images. A whole supply industry is based on its use. The big china clay industry has three principal customers. They are ceramics, paint and paper. China clay is used extensively in the manufacture of paper, and the paper industry uses around 50% of the supply, which is clearly of great importance to the china clay industry. It is now losing some ground to calcium carbonate, which is a loading material available inexpensively in many areas of the world. However, it cannot be used in an acid system, and this has therefore led to the development of neutral sizing of paper, which is particularly helpful when manufacturing papers for which good ageing properties are seen to be important. A third important group of loading materials conferring opacity on the product, is titanium dioxide, which, although not widely used, features in the manufacture of papers that need to be highly opaque, such as laminating grades for furniture and for kitchen working surface applications. Its price and therefore its utilisation are quite different in different parts of the world, being, for example, more expensive in Europe, where it is used only sparingly, and a little less expensive in the United States, where its use in small amounts is a little more general.

It is interesting to have some idea of the relative costs and amounts used of these minerals, as follows:

- China clay, 45%, US$200/tonne.
- Carbonate, 45%, US$200/tonne.
- Talc, 5%, US$300/tonne.
- Titanium, 5%, US$1000–1500/tonne.

Chemicals and dyestuffs

Nothing has been said so far about the other chemicals used in the paper manufacturing process to improve the product. Most papers need to be treated to improve their printability, and so that they will receive ink on their surface. Starch, natural and synthetic size are all used in quantity to promote these properties, and starch and latex are used to promote the adherence of mineral coatings to paper. Dyestuffs, both optical dyes for producing enhanced brightness, and conventional dyes for colouring the paper, form another important aspect of the chemical industry serving the paper industry, as does a variety of other materials such as retention aids, slime treatment chemicals, washing materials and alum. The list is long. These chemicals are important to the pulp and paper industry, and their consumption adds a sizeable annual value to the industry. Chemical costs in the product will be around US$30–40/tonne and this grosses to a global sum of around US$10 billion spent on chemicals consumed annually in paper manufacture and US$5 billion in the pulping sector.

This subject will be dealt with more fully in Chapter 6.

Converting

It is difficult to produce a clear and concise definition of the term 'converting' as applied to the paper industry. Perhaps the best description of conversion is the treatment of paper by further processing after it has been reeled up on the paper machine winder. Some conversion takes place on a large scale within the paper mill, such as cutting sheets. Some conversion has become a business in itself, such as the manufacture of corrugated paper by laminating fluting medium to liner which is then used to make the familiar corrugated case. Some converters can be small one-person businesses operating, for example, a single label stamping and printing machine in a garage.

This section describes just a few of the major converting processes and plants that have grown up around the paper industry to give a feel for the very large and diverse converting industry based on the use of paper. They are all multimillion dollar industries in their own right, and a small selection is briefly discussed here.

Those considered are:

- Reel conversion to slit reels and sheets.
- Corrugating manufacture.
- Envelope manufacture.
- Self-adhesive label paper production.
- The photographic paper industry.
- Newspapers.

- Laminate industry.
- Merchanting.

Reel conversion to slit reels and sheets

Most paper mills run a finishing operation, but some trade finishing operations have now grown up where purchased reels are dealt with. Among the trade finishing processes carried out are the following:

1. *Rereeling*. Here defective paper is removed from the paper machine reels, and joins properly made. After this the paper will continue down the process line to more conversion machinery.
2. *Slitting*. Here paper is edge trimmed, cut to width and reeled up to a selected length.
3. *Supercalendering*. Here the paper is treated on a machine comprising a stack of heated rolls loaded on the ends, so that when the paper passes through the roll nips, it is smoothed and polished and the printing surface enhanced. A similar process is embossing where a pattern is pressed into the paper.
4. *Sheeting*. Paper here is cut several reels at a time to sheet size. It may be possible to cut two different sheet sizes at the same time. It may also be possible to operate a sheet reject system, detecting faults – predominantly joins – by sensor. Sheet counting may be carried out automatically on the slitter. More specific machines may be used to cut small sheet sizes. Ream wrapping may also form an integral part of these production lines.
5. *Wrapping*. Sheet and reel wrapping may take place automatically, as may labelling.
6. *Counting*. If counting is not integral in the sheeting process, it may be carried out at a separate sheet-counting station prior to wrapping.
7. *Palletisation and storage*. Palletisation forms a convenient way of stacking sheets for storage and despatch, and reels may be stored in a similar fashion. Technical papers are now being given at conferences, addressing the whole question of the automatic warehousing of cut sheets and reels.

To set up a modern finishing line is a project costing several millions of dollars. It will take product from the paper machine and carry out all the above processes where relevant, including also packing, bar coding, wrapping and storing or dispatching paper at the rate the mill is producing – say, for example, 1500 tonnes per day.

Such an operation needs to have the correct equipment and

information systems. It needs to be well automated to control labour costs, and it needs to be efficient to enable customers to run 'just in time' systems without the mill keeping money tied up in vast stocks. The means of carrying it out have already been described, and it is clear that the equipment involved in a present-day conversion plant of this sort is impressive. Besides the finishing operation in a paper mill, the work may also be carried out in specialised factories, located in the centre of their markets. These will have similar equipment to the mill. The scale of equipment cost will be on a much lower level than that of paper production equipment, but will nevertheless be of the order of US$30 million.

Corrugating

A corrugating factory uses corrugating medium and liner board, supplied defect-free, in reel form. These grades will be fed through a machine which will flute the corrugating medium, and glue it on both sides with a starch-based adhesive to the liner board. The board may then be either rolled up and cut to suitable lengths as a wrapping material, or, more usually, be sheeted to ordered sizes from which cases will be stamped. The factory will often have a design department which will design cases to suit particular applications.

This is a major industry that is growing quickly. The machinery is specific and complex. In 1994 65 million tonnes of paper were converted world wide. This represents a quarter of all the paper produced in that year.

Envelope manufacture

Modern envelope machines are reel-fed, and produce envelopes at a very high rate. In one process they can cut blanks, which may contain windows which they gum and seal to form the envelope, and then gum flaps with either moistenable or self-adhesive gums. They then count and box the product. The envelope may or may not be personalised with a printed message or logo, and it may or may not have an internal opacifying print.

The machinery is specialised and produced by a limited number of envelope machinery suppliers.

Self-adhesive label paper (SALP)

This industry uses principally two varieties of paper as raw material. One type is paper suitable for silicone coating, for example a glassine-type paper, while the other is suitable for use as a face label paper, for example an art glossy paper, to be used to make high-quality printed labels, or paper for a computer listing grade, used for sticking contents information on a box.

The manufacture of SALP is carried out on a complex coating machine. Glassine or a similar base is coated sparingly and carefully with expensive silicone and dried. An adhesive is coated on top of the silicone, then dried and married up with the face paper to which the adhesive sticks. On subsequent separation in use as a label, the adhesive has stuck to the face paper, and the silicone-coated paper is stripped off and is thrown away, leaving the adhesive-coated label exposed.

The production line is again capital-intensive. It has fast become one that makes a commodity grade, and it is usually operated 24 hours a day, seven days a week. Subsequent processing is similar to a converting line in a paper mill. The machine reels may be rewound, and slit to narrow customer length sizes, or sheeted, guillotined and packed. It is not unusual to have a plain label converting line on the site, when the output will include reels, or small sheets of unprinted customer labels.

After this, some of the product may be converted on site into printed labels or more often sent to specialist label makers, whose businesses vary in size from a small single machine operation to the operation on a major factory complex, operating many label printing machines and turning out many qualities from sheets of plain labels to the most sophisticated of colour-printed labels processed on four- or six-colour presses.

The photographic paper industry

This is an imaging industry which sets out to produce a light-sensitive coating on a series of substrates which are predominantly film or paper. The paper substrate is made on a paper machine, after which it is either resin-coated (resin-coated base) or coated with inert barium sulphate (fibre-based paper).

The next processes are carried out at the sensitising factory, under total brown or red safe lighting, depending on the light sensitivity of the coating. There the paper is coated with emulsion and gelatin, and dried relatively slowly on a long coating machine, and reeled up. It is then slit to remove defects and cut to customer widths and lengths, or it is sheeted and packed. All of these operations take place in the near dark, and the slitting equipment is some of the most sophisticated in the market. It is only when the product is wrapped with a light-tight wrapping that it can emerge into the light.

Quality is of vital importance to the product, and the machines are carefully designed to produce a blemish-free product. Machine-induced blemishes can often only be found after the product has been sampled, coated and developed, and then carefully inspected,

The industry, including the production of light-sensitive film, is an important business.

Newspapers

The newspaper industry is clearly dependent upon paper, where speed of newspaper printing and runnability, and therefore precise drive control of machinery, are paramount. The paper must be capable of being run at speed without fault over the printing presses. It must also be able to accept printing ink, and four-colour printing. The machines are sophisticated printers and dryers, and they are put together into one process in such a way as to produce newspapers very fast. They have the complexity of paper machines, and are controlled by costly and sophisticated sectional drives which allow excellent speed control, to contend with very rapid machine acceleration and deceleration.

The laminating industry

Once again the industry is totally dependent upon reels of paper. In some cases the paper is laminated to a chipboard substrate, and in others the surface paper is laminated to several layers of core papers. Thus there are two types of paper involved. The surface paper must be opaque, and capable of being printed in very sophisticated and exact patterns. The core papers are very absorbent strong papers. The process of making laminates is to impregnate them heavily with resin, and to press them in big presses at high pressure so that the laminate is formed. Much of the product is then used by furniture and DIY outlets.

Merchanting

Paper merchants buy stock from paper mills for resale to the consumer. They have a business dealing with a wide number of grades in various sizes, usually of office and printing papers, and they are produced 'off the shelf' for immediate delivery. They may also keep customer-specified grades if the stock lines are insufficient. To accomplish this, merchants require a knowledge of the production cycles of the mills from whom they buy so that they can ensure that their stock lines are maintained without over-ordering and so locking up working capital.

Merchanting is an international operation, bringing the product to the end user. The merchant has the responsibility of providing a satisfactory, fast distribution network. This will often manifest itself through distribution organisations and agencies world-wide. Merchants are often owned by paper-manufacturing companies.

Summary

This chapter has described in outline the processes involved in the manufacture of pulp and paper. The engineering aspects of the industry have been seen to be complex, capital-intensive and advanced. New plant comes on stream, bringing with it the necessity to sell large amounts of new material. Machines 10 m wide operated at speeds of near 2000 m/min and output rates of over 250 000 tonnes per annum are to be expected. Raw materials other than fibre used in papermaking have been briefly described, but they will be more fully dealt with in Chapter 6.

Some converting operations have been outlined, as paper is of vital importance to many converting operations.

CHAPTER

3

World consumption

Detailed information is presented about the world consumption of paper linked to world population and world per capita consumption. The information is provided in a comprehensive series of figures spanning 20 years, which shows the growth in terms of volume consumed, and some of the implications of this growth. The chapter deals with the information about paper globally, before considering it in a grade-by-grade analysis. It prepares the ground for a later discussion about the ways in which the consumption can be satisfied, through production, import and export. The summary indicates the grade-by-grade growth of paper world-wide, indexed to 100% in 1975.

Introduction

A wealth of statistics is available which deals with the consumption of paper across the world. The approach taken in this chapter is to digest present world statistics in the form of easily assimilated figures. World consumption is easily defined and simply stated. Figure 3.1 shows total paper consumption spanning the past 20 years, increasing from 127 million tonnes in 1975 to 276 million tonnes in 1995. The figure illustrates the magnitude, and indeed the importance of the international trade.

Globally the growth rate (Fig. 3.2) over the past 20 years has been at an average of 5.9% per annum, compared with the reference point

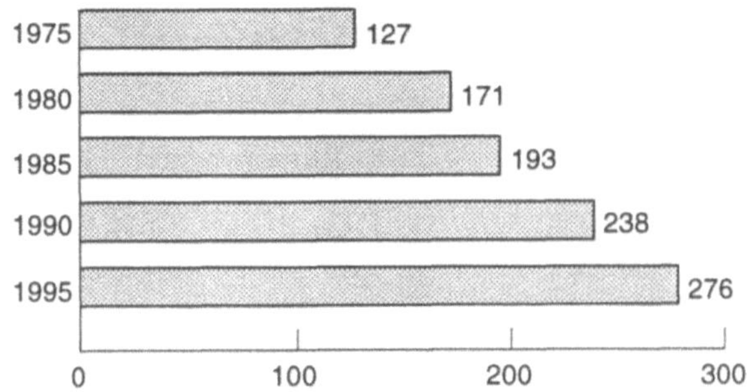

3.1 Total world consumption of paper (million tonnes).

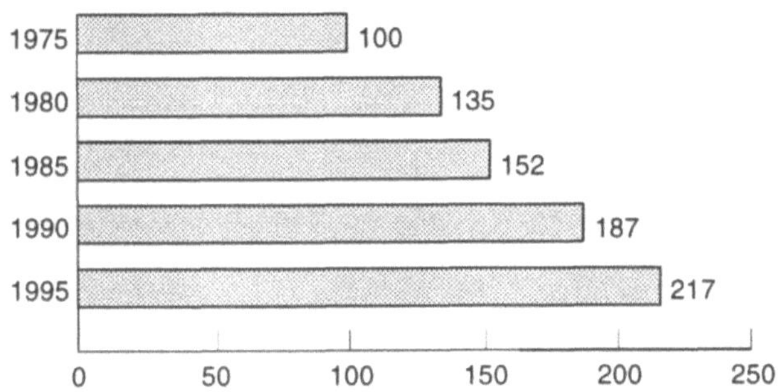

3.2 World consumption of paper indexed to 100 in 1975.

in 1975. However, the rate of growth is reducing with time. The growth for the ten years from 1975 to 1985 ran at 5.1% per annum, while the growth over the next five years from 1985 to 1990, taking the 1985 figure as the reference for comparison, clearly slowed to 4.8% per annum. In the last period, from 1990 to 1995, the average consumption showed a further reduction of growth to 3.2% per annum in the recession years from which the world is emerging. When growth for the next decade is considered, many observers choose the arguably conservative figure of 3.0% per annum.

Paper sells at a wide variety of prices but if one considers an average selling price to be around US$1200 per tonne, the value of the paper sold is around US$330 000 million each year. However it is viewed, consumption in the industry is growing by at least 3.0% per annum, which is a most healthy state of affairs.

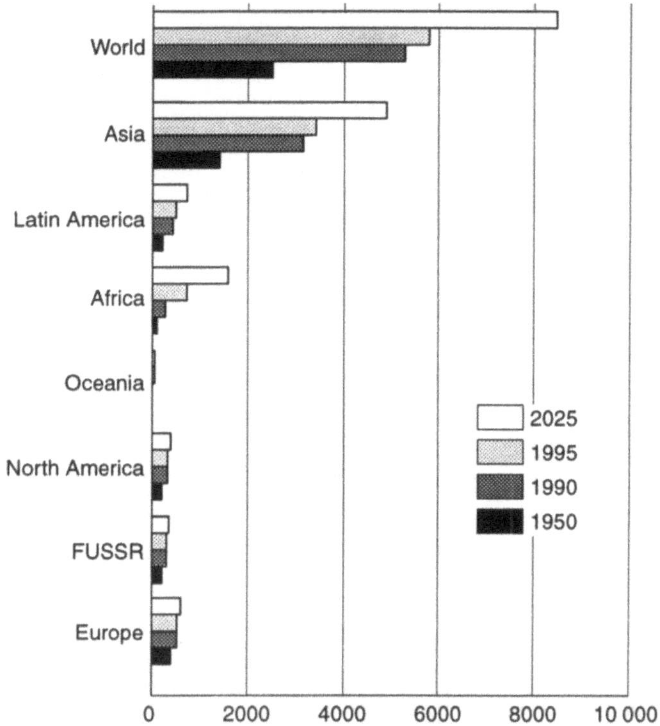

3.3 Growth in world population in millions 1950–1995 and estimated to 2025. (Note: FUSSR = Former Union of States of Soviet Russia)

Link to population and per capita consumption

Population

World population must clearly have an influence on paper consumption. Figure 3.3 shows the increasing population from 1950 taken to an estimated level in the year 2025, and highlights the well-known facts of population explosion and growth. The major centres of population are Asia (and China is significant within Asia), Africa and Latin America.

The figure for 2025 shows the large numbers in Asia, estimated at 4900 million, of which China accounts for 31%, with a population estimated at 1500 million. Africa shows figures of the same order of magnitude as China, and the growth rate in these continents is high,

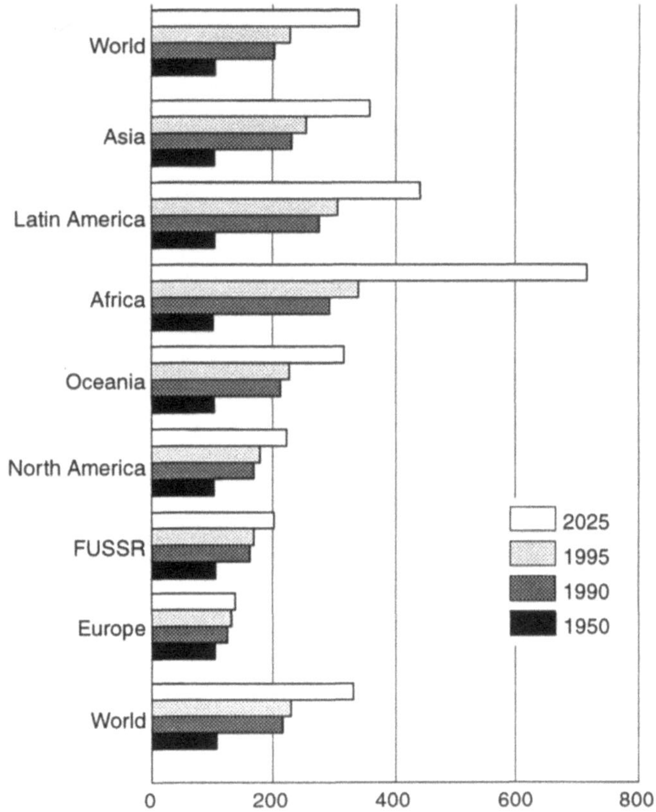

3.4 World population indexed to 100 in 1950.

more than doubling in Asia and Africa and with a slightly lower rate of increase in China over the period. All of these growth rates are in excess of the remaining continents.

Figure 3.4 shows the world's population in the major regions indexed to 100 in 1950, and estimated to the year 2025. In the span of the 75 years under consideration, world population is estimated to increase by an average of 3.2% a year. The growth rates are different in different continents. Africa (8.2%), Latin America (4.5%) and Asia (3.4%) are growing above the average rate, while China, within the Asian statistics is growing at 2.4%, faster than Europe (0.5%) and North America (1.6%) but slower than the world average.

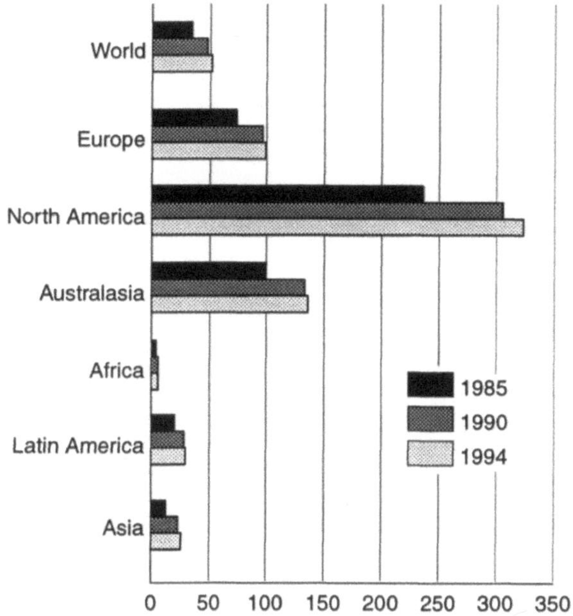

3.5 Per capita consumption of paper by region (kg per capita).

Per capita consumption

As we have seen, paper consumption is considered by many to be an indication of the degree of literacy and sophistication of a country. The statistics for actual paper consumption in kilograms per capita in the regions of the world today are shown in Fig. 3.5. The histogram indicates the very big differences in the annual per capita consumption from the North American figure of 321.6 kg per capita to the African 5.2 kg per capita. The Asian figure of 23.6 kg per capita includes a figure of 20.2 kg per capita for the People's Republic of China. The USA, with 333 kg per capita, leads the way, with Japan at 231 kg per capita and Western Europe with 173 kg per capita. India (2.7 kg per capita) also shows very low per capita consumption.

Taking the USA from the North American figures, it can be seen that, with a population of 250 million and a per capita consumption of 333 kg, it manages a paper consumption of 86 million tonnes per annum, while China, with a population estimated to reach 1500 million in 2025 and a per capita consumption of 10 kg consumes only about 24 million tonnes. One can foresee a healthy growth of paper consumption in China, over the next decades. The same observations

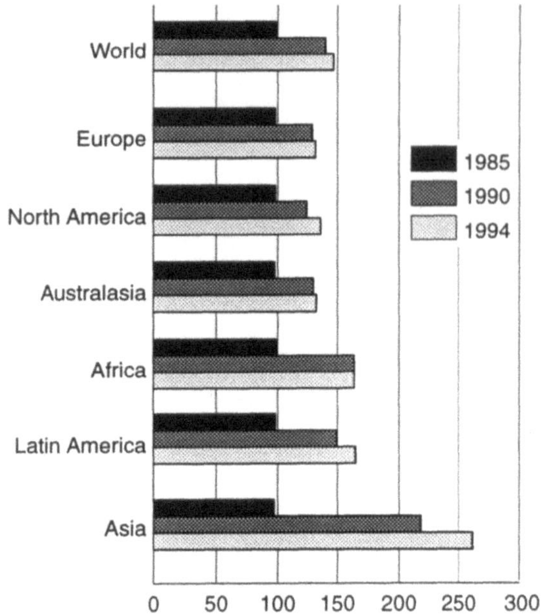

3.6. Per capita consumption of paper by region indexed to 100 in 1985.

apply to India and Africa, although the growth will probably be much slower.

Indexing these figures to 100 in 1985 (Fig. 3.6) shows per capita growth above the world average in Africa, Latin America and Asia, and below the world average in North America and Europe. The recession of the early 1990s has clearly influenced this position, showing in most cases that the increase from 1990 to 1994 was slower than the corresponding period from 1985 to 1990.

World consumption

Having taken global figures of paper consumption, and seen them alongside population growth and per capita consumption, it is now appropriate to consider figures of total world consumption and consumption by region. Consumption is seen in Fig. 3.7 to be highest in North America, followed closely by Asia and Europe. Lowest consumption is in Australasia, where the population is small, and Africa, with Latin America only a little higher.

As Fig. 3.8 shows, the largest percentage consumer is also the North American continent. It commanded 40% of the world's consumption in 1975, the figure falling by 2% in the ten years to

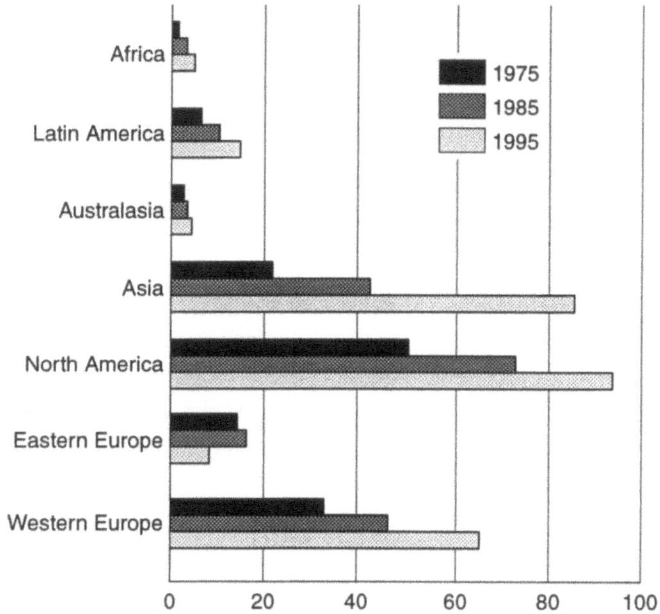

3.7 Paper consumption by region (million tonnes).

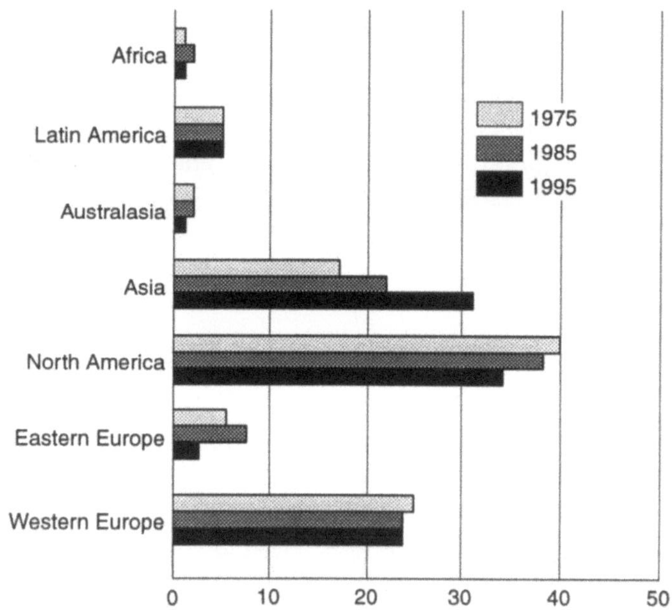

3.8 Percentage of world paper consumption by region.

1985, and a further 2% to 34% in the next nine years to 1994. While this was taking place, the Asian figure increased from 16% in 1975 to

Table 3.1 Factors influencing world consumption of paper (1994)

	World	North America	Asia	West Europe	Latin America	East Europe	Africa	Australasia
Population (million 1990)	5669	293	3365	385	473	415	710	27
Per capita (kg per capita)	49	322	25	172	31	19	3	153
Consumption (million tonnes)	276	94	85	66	15	8	4	4

22% in 1985 and then further to 30% of the world's consumption by 1994. It is now poised to overtake North America as the region consuming the most paper annually.

Western Europe maintained a steady 24–25% throughout the period, but Eastern Europe fell from 11% in 1975 to 8% in 1985 and then further to 3% by 1994. As the Eastern European economy improves one will expect its share of the world consumption once again to be more buoyant. Africa, Australasia and Latin America only have a small share of world consumption.

The loose relationships between total consumption, per capita consumption and population are brought together in Table 3.1 drawing attention to the potential markets in Africa, Latin America, Eastern Europe and Asia.

Having considered the global paper scene, a breakdown worldwide of the major grades is now given.

Consumption by grade across the world

Figure 3.9 shows the world consumption statistics for the major paper grades. The consumption of printings and writings, and of corrugated case materials both show figures for 1994 in excess of 50 million tonnes. Newsprint, packaging and tissue grades are also substantial representatives of paper consumption. Tissue is perhaps the most remarkable when one considers the very light-weight nature of the product – about a half to one-third the weight of newsprint, on a grams per square metre (g/m^2) basis. Tissue weighs around 21 g/m^2 while newsprint is in the range 45–60 g/m^2.

Indexing these figures also shows some interesting results (Fig.

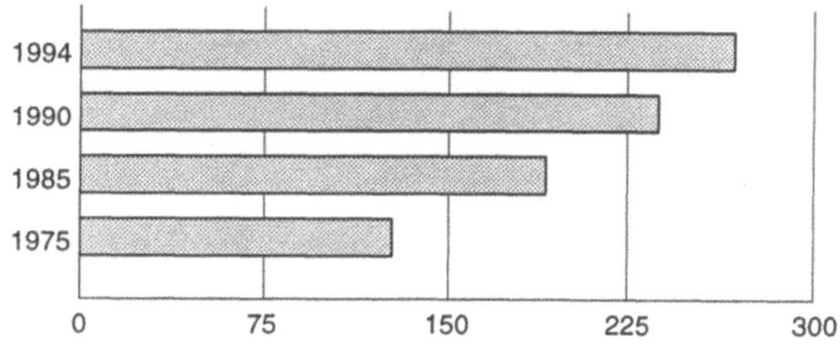

(a)

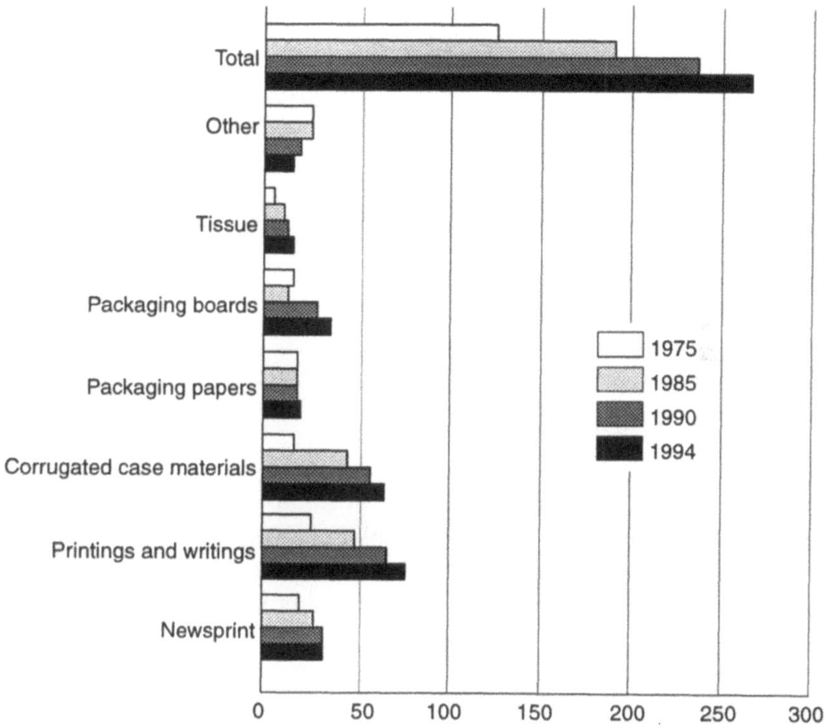

3.9 (a) World consumption of paper (million tonnes); (b) World consumption by grade (million tonnes).

3.10). The order of growth of each product, from 1975 to 1994, ranks them in the order corrugated case material, tissue, printings and writings, packaging boards and newsprint. It is interesting to note that there is virtually no growth in packaging papers, which have apparently been losing market share to plastic-based products, and there is also a distortion in the figures from reduction of grammage of packaging papers. It is interesting to speculate on whether or not this will continue when the cycle of opinion focuses once more on the conservation of oil-based products.

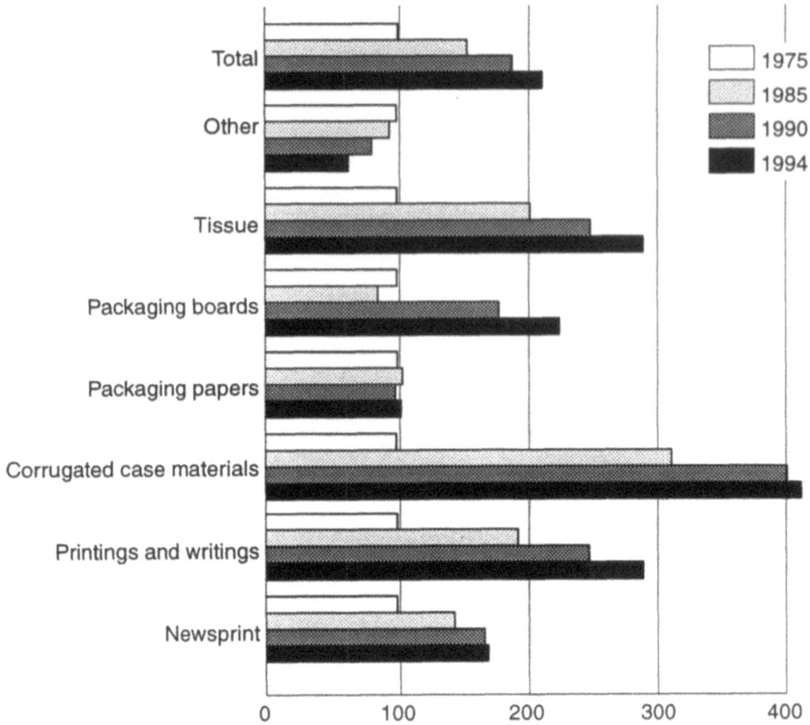

3.10 World consumption by grade indexed to 100 in 1975.

So far the discussion has been about the use of paper generally on a global basis, but it is necessary now to consider the categorised consumption of the principal paper grades across the world. They were classified in Chapter 1 in the seven main groups of newsprint, corrugating materials, packaging papers, boards, printing and writing grades, hygienic tissue grades and other. We now consider information contained in the analysis of each sector. The following sections show the growth of each of these groups of papers since 1975. As a basis for comparison, the figures in Fig. 3.11 and 3.12 show tonnage consumption and consumption indexed to 100 in 1975. World consumption achieved 268 million tonnes in 1994, at an average growth rate over 20 years of 3.1% per annum.

Newsprint

Newspaper consumption (Fig. 3.13) has moved from a number of years of contraction of demand in some countries, for example in the UK, in the period up to 1975, to its resurgence in the 1980s. It has also

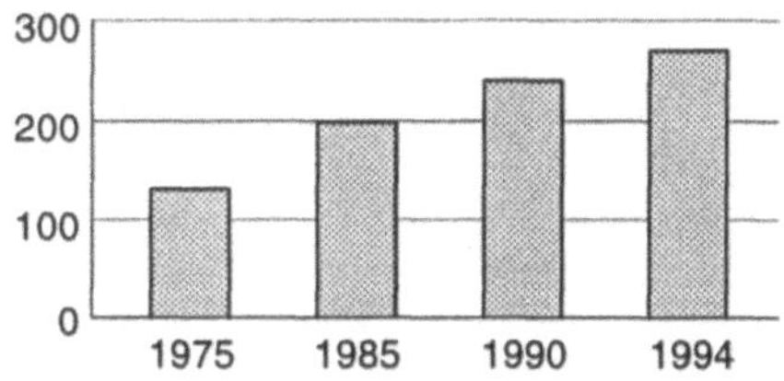

3.11 World consumption of paper (million tonnes).

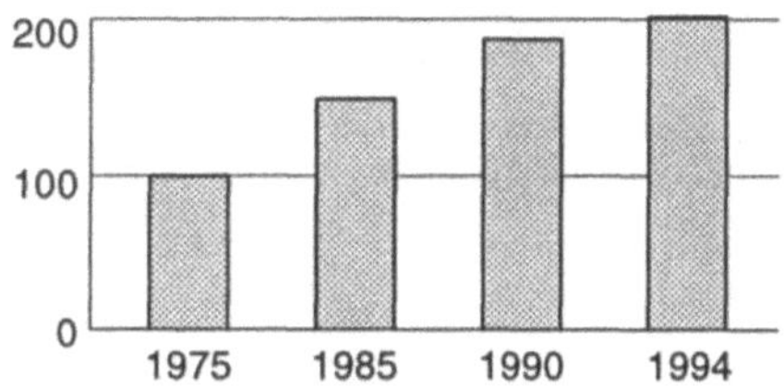

3.12 World consumption of paper indexed to 100 in 1975.

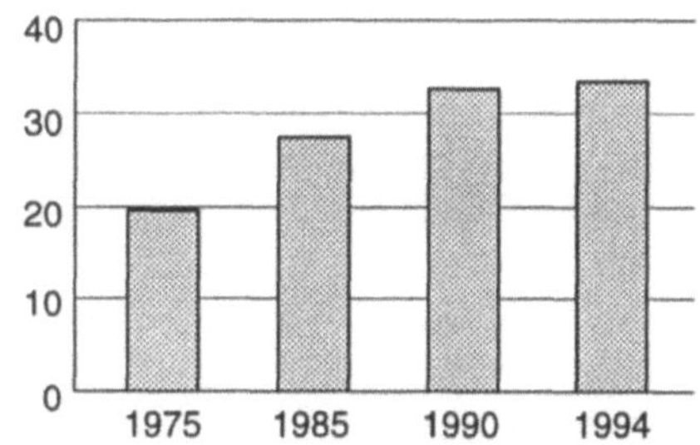

3.13 World newsprint consumption (million tonnes).

changed from complete reliance on wood pulp, to, in some cases today, manufacture from up to 100% recycled newsprint, as in the case of SCA Aylesford. This is an interesting development since the traditional reason for siting newsprint mills near forests and energy sources, in the form of hydroelectric power, is no longer valid. At the same time, there has been a great improvement in the product so that it is now suitable for printing in colour at high speed. There have also been huge engineering developments so that the paper can be produced wide (10 m) and fast (speeds up to 2000 m/min) on new machines and in some cases, in new mills.

Demand has now moved towards the lighter range (45 g/m^2) of newsprint basis weight owing in no small part to the use of postal services for some distribution where weight is important – reminiscent of airmail editions being produced on more expensive airmail paper – and printability is of ever-increasing importance. The demand

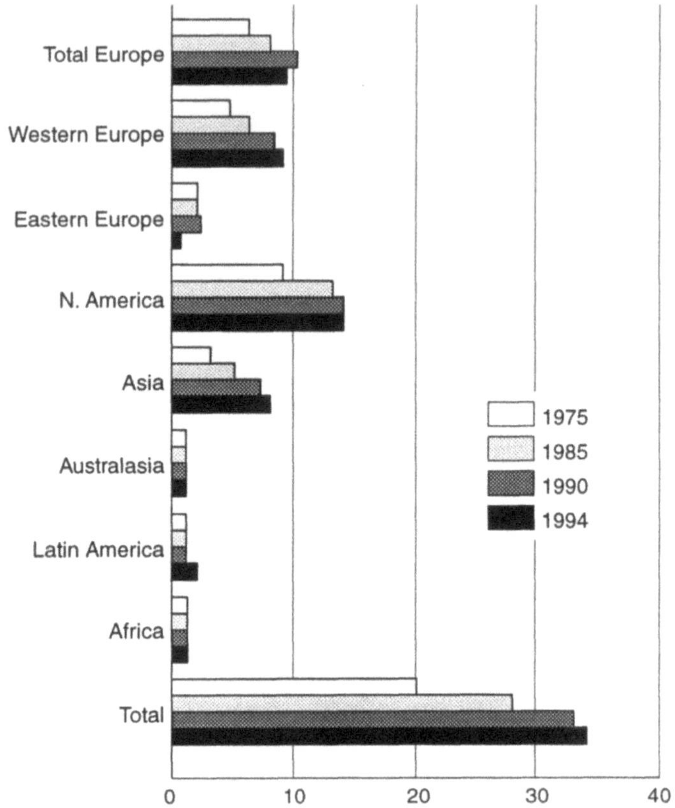

3.14 Newsprint consumption by region (million tonnes).

figures have not shown a consistent growth, owing in part to this reduction of weight - one would assume the area produced to have shown the expected growth, but again the growth figures have also been influenced by the world recession, and the reduction in advertising carried at the time. The commodity is in some ways getting closer to light-weight coated papers, which are more expensive.

The grade has a high strategic and political importance. Newspaper publishers are responding to pressure from the public to join the environmental debate by declaring the average percentage of recycled fibre from which their newspapers are manufactured.

Figure 3.14 shows that North America is above all the major consumer, followed by Europe, which is broken down into Western and Eastern Europe. This breakdown points to the unsettled nature of consumption in Eastern Europe, and the opportunity for growth. It is worth having a closer look at the major consumers in Asia which follows North America and Europe (Fig. 3.15). Japan is a major

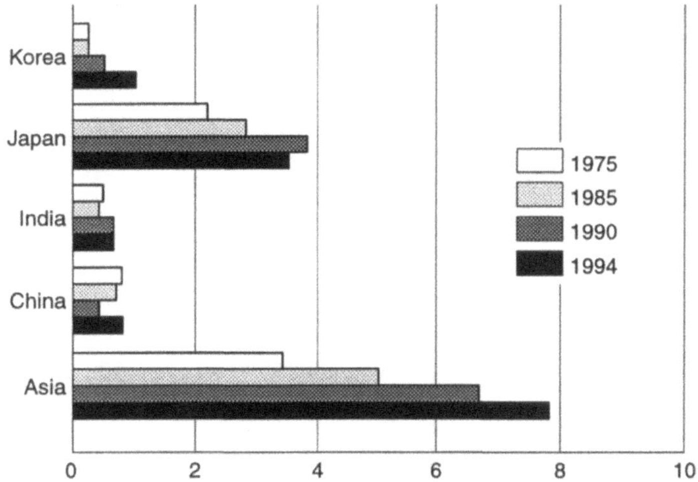

3.15 Newsprint consumption in Asia (million tonnes).

consumer of newsprint, using some 3.5 million tonnes in 1994, even though the figure fell back slightly from the 1985 value of 3.8 million tonnes. Korea is close to using a million tonnes, and China and India are clearly likely to increase their consumption substantially.

Printing and writing papers

Since 1975, the growth in printings and writings (287%; Fig. 3.16) has with tissue (290%) been the greatest of all of the sectors except corrugating materials, since the datum in 1975, taken as 100%. This was shown in Fig. 3.10. Taken over the whole of this period of time, printings and writings have grown at an average of 9.8% per annum. Although slowing a little in the last few years under consideration, growth figures remain impressive and after 3.1% in 1993, they showed 7.7% in 1994.

It is one area of the paper business that has experienced the greatest change since the 1950s. Carbonless paper was not in production until the 1950s when it was introduced as a speciality grade. Today its volume has grown significantly, and it is now viewed as a commodity grade. Printings and writings were often bespoke small tonnage printing papers. Those markets have now become uneconomic for paper made on old, slow machines. The sorting operation which employed hundreds of paper sorters has become automated. The mills now use sophisticated process control and fault detection equipment, and the finishing department of a mill, while

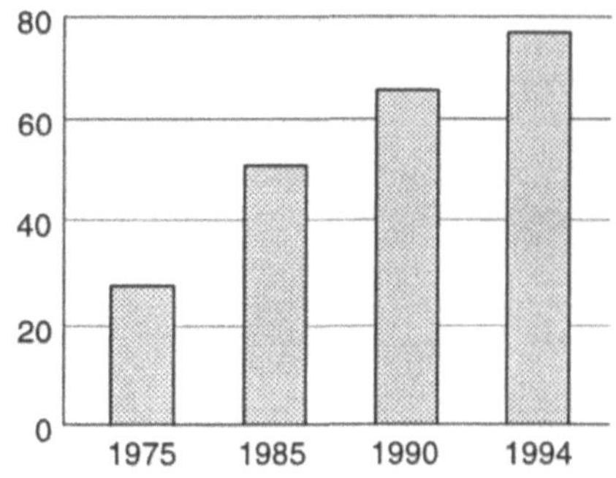

3.16 **Printings and writings consumption (million tonnes)**

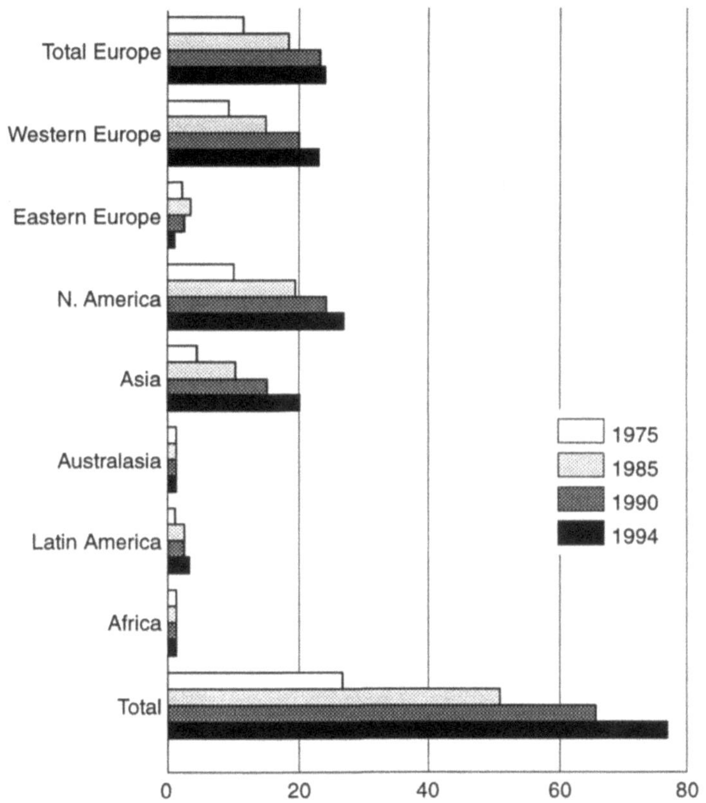

3.17 **Printings and writings consumption by region (million tonnes).**

being the most labour-intensive, nevertheless still shows a reduction of its work-force by up to as much as 90% in a period of 30 years. A medium sized mill, manufacturing 50 000 tonnes per annum, will today probably have a work-force of no more than 150 to 200 people. Output from machine rebuilds has risen, and specialist machines are being laid down for the production of light-weight, usually coated, printings.

However, the impressive growth in the sector consumption figures, masks the demise of other high-volume grades such as continuous stationery and some copier papers.

Considering now the world scene for this grade (Fig. 3.17) we see again the impressive consumption in Asia following closely behind Europe and North America. There has been a steady growth in consumption across Europe, North America and Asia. Some 84% of this growth in Asia is accounted for by Japan (49%), China (22%), India (7.5%) and Korea (7%).

Board consumption

The packaging board segment of the industry was severely affected when the industrial changes of the 1970s were taking place. Huge volumes of tonnage were lost when major producers closed down, or dramatically reduced their output. Stability is just returning to this sector of the market, and after consumption fell until 1985, it has now increased by 20 million tonnes to 36 million tonnes, from a base of 16 million tonnes in 1975 (Fig. 3.18). The product needs to be functional in providing protection to the packed product besides having a sales appeal in promoting the contents of the carton. Consequently multi-ply boards are most usual, with the outside ply often coated to impart printability, and the inside ply coated to impart special functional properties such as grease resistance. Thus it can be seen that the qualities span the entire range from a container board to a sophisticated highly glazed printable grade for prestigious work in, for example, the sophisticated field of tobacco and cosmetic packaging.

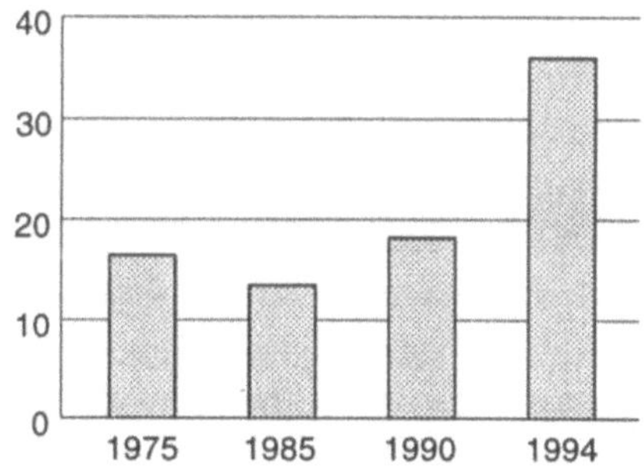

3.18 World board consumption (million tonnes).

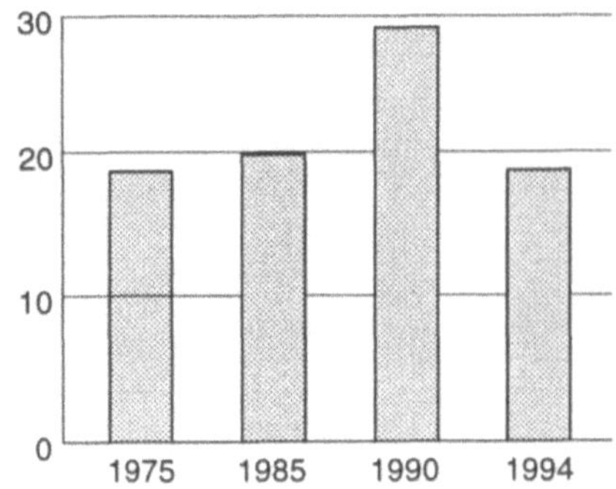

3.19 World packaging paper consumption (million tonnes).

The grade also makes an important environmental contribution in that the major raw material component in the production of this sector is recycled waste.

Packaging paper consumption

This sector of the industry has experienced a great deal of competition from non-paper materials over the past two decades. The paper market has been led by integrated mills which were able to produce superior products - sack kraft and unglazed kraft - at very competitive rates. Environmental pressure has led to a grammage reduction - known in the States as 'Liteweighting'. Low selling prices, combined with the use of plastic materials as alternative products, contributed to the dismal growth from 1975 to 1985 when the figure was a bare 0.1% per annum (Fig. 3.19).

Corrugating case materials

After the somewhat depressing story of packaging boards and papers it is interesting to come to the success story provided by the corrugated case materials sector. Even in the rationalising 1970s the growth did not stop, and it is now showing extremely good growth at 450% of its 1975 level (Fig. 3.20). With a 6.5% growth in 1994 this demanded an annual increase of capacity of around 2 million tonnes or an average of 1.3 million tonnes per annum over the 19 year period.

The product is assembled from three paper products: the inside corrugating medium, the backing sheet and the liner which can be a solid board and can carry printed decoration. Besides being key to the

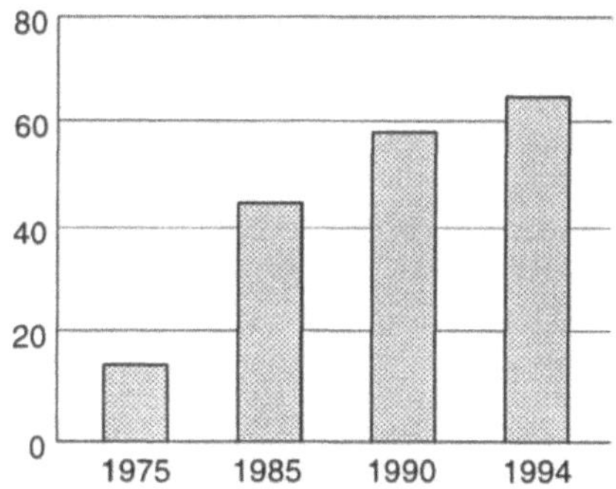

3.20 Corrugating materials consumption (million tonnes).

paper industry, corrugated cases have also spawned a big industry of corrugated box makers, and they have their own corrugated container associations, exhibitions and all the accoutrements of a major industry in its own right.

The corrugated container wins the environmental argument outright. It uses primarily waste paper as a raw material, it is a major contributor to the environmental credentials of the paper industry, and it can be easily recycled.

In 1994 the biggest consumers in the world (who are also almost always the biggest producers) each used more than 2 million tonnes:

- North America - USA (25.8 million tonnes).
- Europe - France (2.8 million tonnes), Italy (2.6 million tonnes), UK (2.3 million tonnes).
- Asia - Japan (8.8 million tonnes), Taiwan (1.9 million tonnes), Korea (2.0 million tonnes).
- South America - Brazil (2 million tonnes).

Hygienic tissue grades

The figures for tissue grades show another strong growth business (Fig. 3.21). Tissue consumption has increased steeply since 1975 averaging out at 9.6% per annum. Products developed then are now essentials. Toilet tissue, facial, serviette and towelling grades make up the bulk of the tonnage. The market has swung from being dominated in the 1950s by 'hard' tissues to the now almost complete replacement by 'soft' tissues. Tissue implies a paper grammage of around 21 g/m^2, a very light paper and one made on a dedicated paper machine. Because of this dedication, combined with the use of new technology specially adapted for the new soft tissue industry, the machines are fast, not as

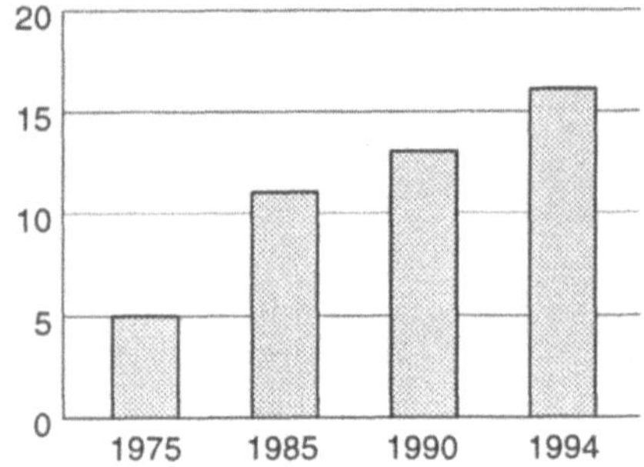

3.21 Hygienic papers consumption (million tonnes).

wide as conventional paper machines and they are less complex and much shorter in length. Steady growth demands a steady investment and this is in fact the position for the tissue sector. The growth is at a rate of around ten new tissue machines each year, which represents a substantial investment, even though the machines are much simpler and less expensive, and have significantly lower output than, say, a newsprint machine. The papers are in a sensitive environmental sector of the industry and because of this the companies involved pay a great deal of attention to their environmental images.

Again the USA leads consumption with a steady 5.5 million tonnes, followed by Japan (1.5 million tonnes) China (1.0 million tonnes) and Germany (1.0 million tonnes). In fact the total European market is just over 3 million tonnes, around half the size of the consumption in the USA.

'Other' grades

'Other' grades is the catch-all segment of the industry. The statistics (Fig. 3.22) hide the transference of some grades, for example in the 1950s the highly specialised carbonless copying papers, to the more usual commodity grades as their market has increased, and production has grown to commodity size, but there is nevertheless a good steady tonnage of speciality papers in this sector. It is a sector interesting to the technologist, but less interesting to the economist. Even so, many peripheral businesses depend upon the continuation of these grades. They encompass, for example, cigarette tissue, photographic papers, decorative laminate papers, filter papers, self-adhesive paper stock, tea-bag tissues, cable papers and so on.

There is much dependence upon them by other industry sectors, and the tobacco industry provides a good example of this.

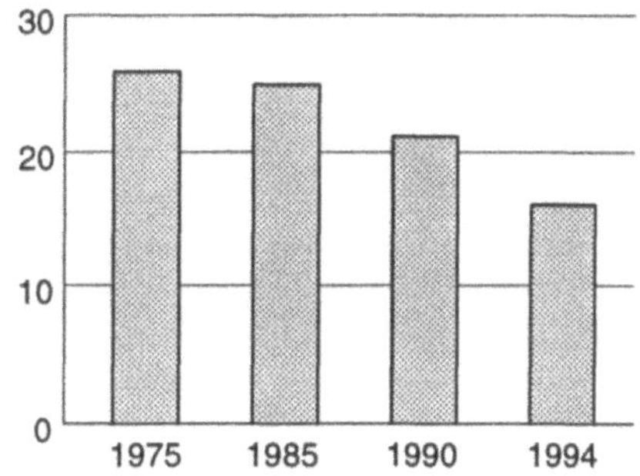

3.22 Other grades consumption (million tonnes).

Summary

Paper consumption will continue to increase over the next decade. In particular, it will respond to increases of population and increasing per capita consumption, and these two factors point to growth in Asia, and China in particular, followed by Africa and India, where the increase in consumption will be above average. Growth in Europe will continue quietly with greater possibilities in Eastern Europe once the political system is established, and there is money available for the regeneration of the paper industry.

Perhaps the best summary is to see the consumption of all principal grades between 1975 and 1994 indexed to 100 in 1975 (Fig. 3.23). By far the best and most consistent increases in consumption are demonstrated by corrugating materials. Following them with about two-thirds of their growth are printings and writings and then tissue. Board is next, just above the average for all papers. Below this average are packaging papers and, perhaps surprisingly, newsprint, although here the picture is a little obscured by product, and particularly grammage changes.

There seems to be every reason why an anticipated average growth of about 3% should be achieved. However, the industry has a cyclic pattern to its growth, and consumption will reflect both need and prosperity.

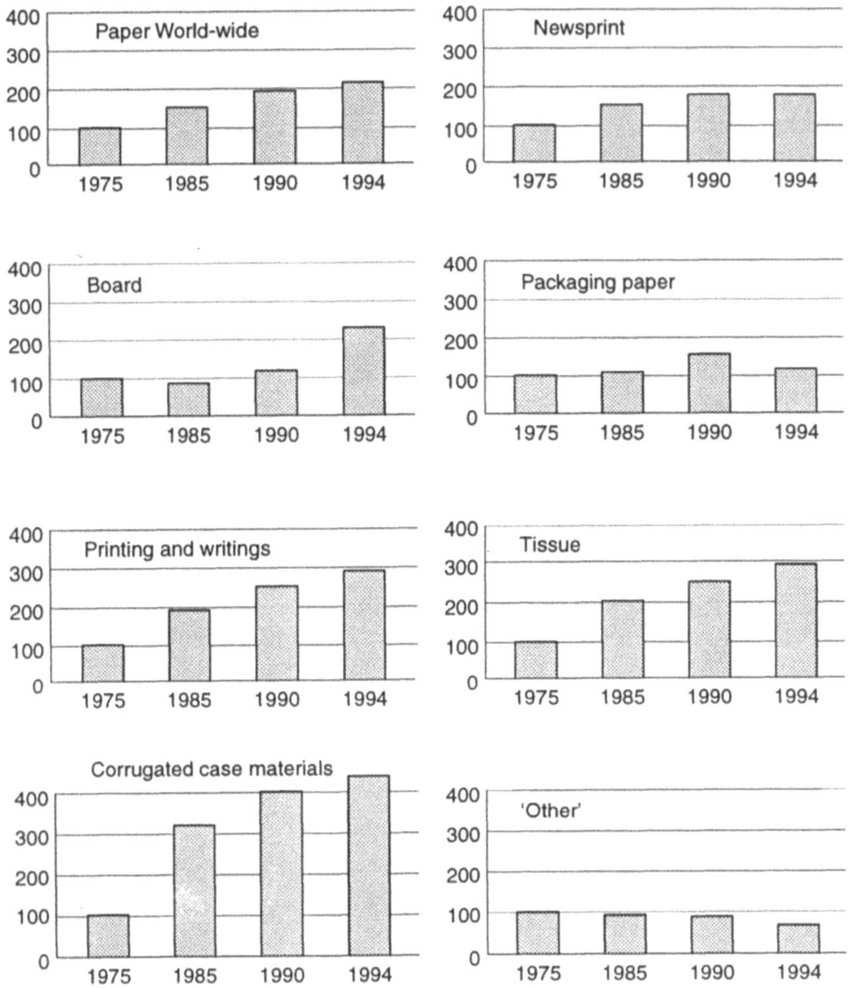

3.23 **Paper growth by grade indexed to 100 in 1975.**

CHAPTER

4

Production, consumption, import and export

The ways in which consumption of paper is satisfied by production, import and export throughout the major regions of the world are now considered. A series of histograms linking these components is considered, and this builds up a picture of import and export across the major regions of the world. The regions considered are Africa, Latin America, Australasia, Asia, North America and Europe. Pulp is dealt with in Chapter 6.

Introduction

Figure 4.1 shows the balance of world consumption, production, import and export from 1975 to 1994. There is nothing exceptional in this figure since the components must be in equilibrium, after allowing for small discrepancies for stock movement. Production must equal consumption on the global scale, and the fact that there is export must be balanced by a corresponding level of import.

The healthy import/export trade across the world of around 70 million tonnes of paper and board is clearly identified in Fig. 4.1.

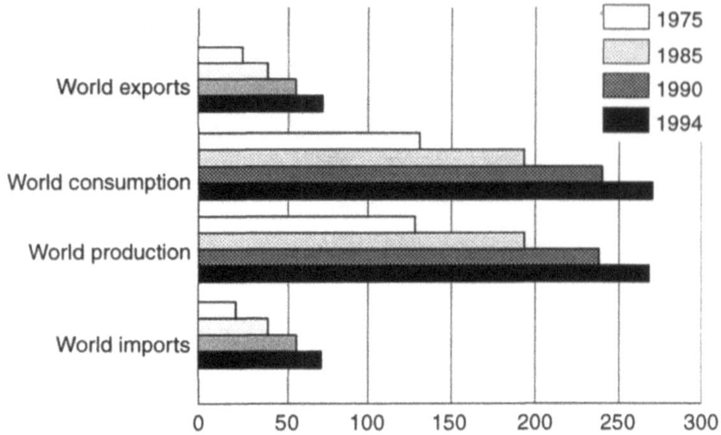

4.1 World paper and board consumption, production, import and export (million tonnes).

Taking an average paper price of US$1200, this puts a crude value on world import and export trade of around US$84 billion. This statistic relates to paper alone. Besides this, there is a pulp volume of 30 million tonnes which, at an average selling price of around US$600, values the import/export trade in pulp at US$18 billion. This brings the value of the import and export trade in pulp and paper to just over US$100 billion. The paper and board statistics (Fig. 4.1) can not be expected to balance exactly since they do not take into account such items as stock adjustments, but the important thing to highlight is the healthy import/export trade.

This chapter will deal with just the paper figures, leaving any discussion of pulp to Chapter 6.

Africa

It has already been seen that the continent of Africa has a large and fast-expanding population. The per capita paper consumption is among the smallest in the world, exceeding 10 kg per person per annum in only two cases, the more important being South Africa, where the figure rises to 43 kg per person. Consequently it is hardly surprising that the paper activity is not very great (Fig. 4.2 and 4.3). Paper consumption is around 3.5 million tonnes and it is supplied by some manufacture within the region and by quite heavy imports which represent around half of the consumption.

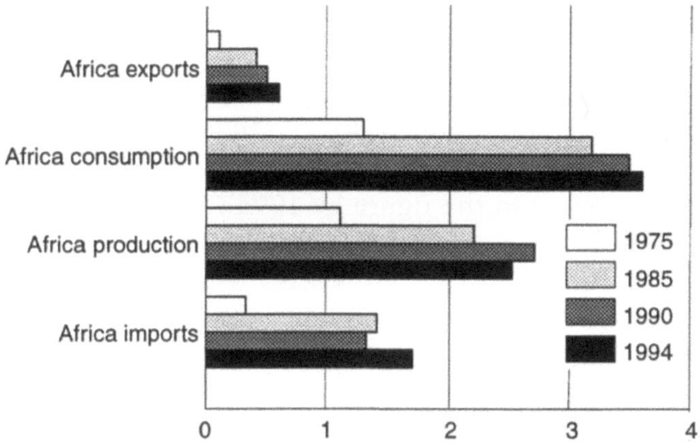

4.2 African paper and board export, consumption, production and import (million tonnes).

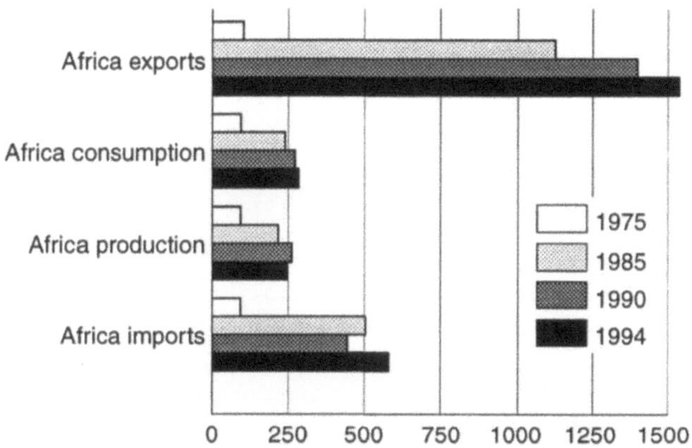

4.3 African paper and board export, consumption indexed to 100 in 1975.

The past decade has shown a steady situation with little growth compared with the substantial growth seen in the period 1975 to 1985. There appears to be a disproportionately high degree of export activity, in fact a doubling, from 1975, but the starting point in 1975 is so low that it barely registers on the histogram, so that this change is unlikely to be significant in world terms. It is unlikely there will be any sudden changes in the forthcoming pattern and this is not seen as an area of immediate growth, even though population is high and the per capita consumption is so small providing the potential for later growth.

Latin America

The Latin American paper consumption grew significantly by an average of 8.3% per annum in the years 1975–94. Exports (Fig. 4.4) were at an extremely low level in 1975 - around 125 000 tonnes, so that the improvement in the figure by 1994 (Fig. 4.5) was an average 80% per annum to 2.1 million tonnes, coming mainly from production capacity installed in Brazil. Considering the next nine years, from 1985 to 1994, which start from a much higher base level, growth was at a level of 18% per annum.

The import, export, production and consumption figures for the region in tonnes are shown in Fig. 4.6, and the figures indexed to 100 in 1975 are shown in Fig. 4.7. They show that Latin America is a net importer, and one of the reasons for this is because its industry is not yet big or diverse enough to keep up with the consumption demand. The histograms demonstrate the leap in imports from 1990 to 1994 to keep up with the increasing consumption. This increase was at a rate of just over 20% per annum.

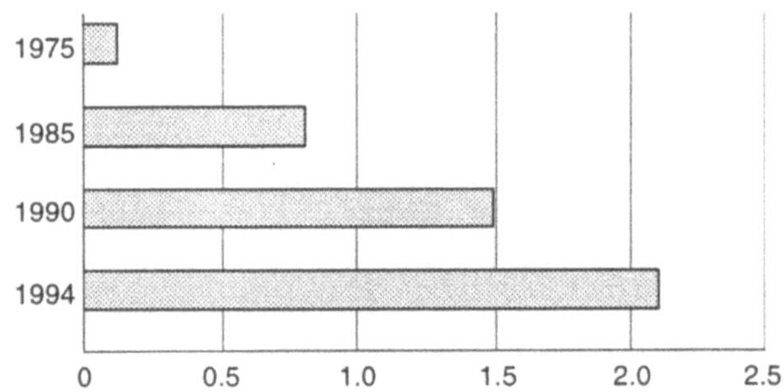

4.4 Latin American paper and board exports (million tonnes).

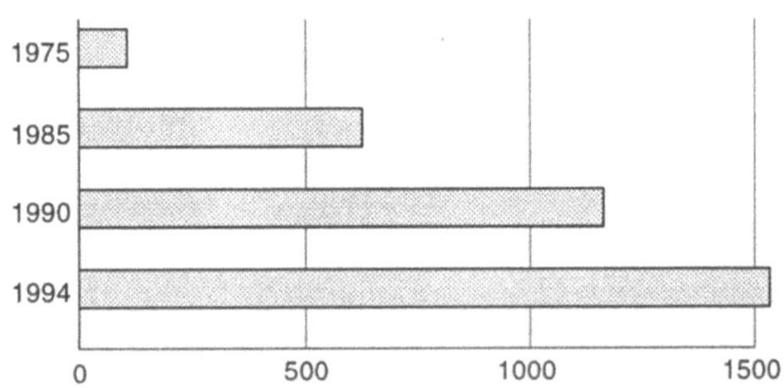

4.5 Latin America paper and board exports indexed to 100 in 1975.

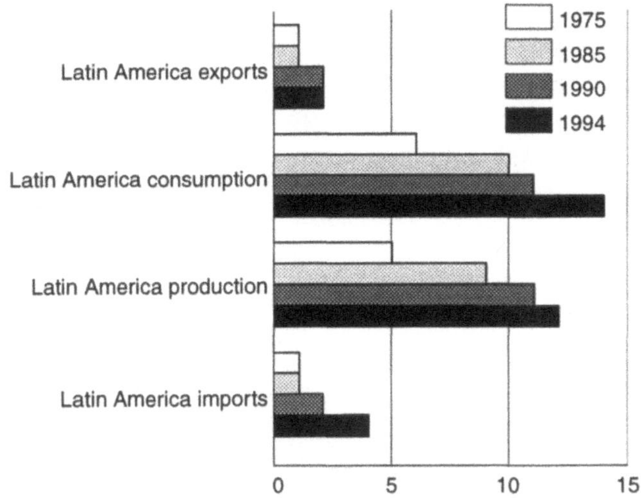

4.6 Latin America paper and board import, production, consumption and export (million tonnes).

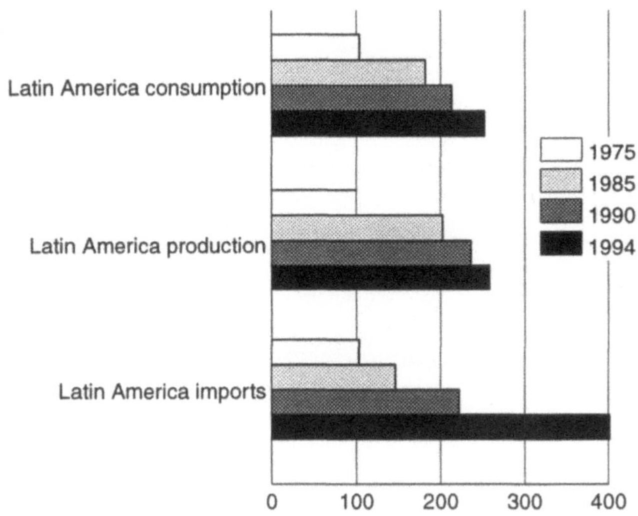

4.7 Latin American paper and board production, consumption and import indexed to 100 in 1975.

In summary, Latin America can be seen as a region where consumption will increase, where indigenous production capacity will be installed, and where the import of paper will continue to increase.

Australasia

The Australasian region is shown as a net importer. The whole picture is one of order and stability. In 1994 the trade was approaching a consumption of 3.7 million tonnes, and with a production of 3.1 million tonnes it imports the balance of 0.6 million tonnes (Fig. 4.8 and 4.9).

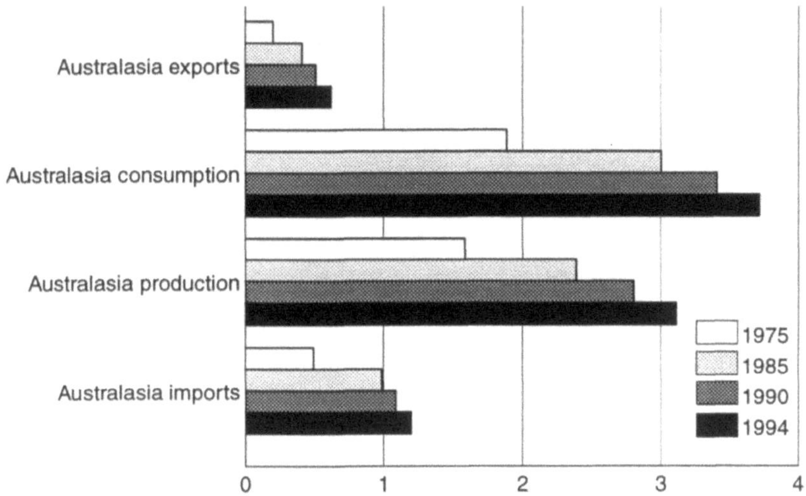

4.8 Australasian paper and board consumption, production and import (million tonnes).

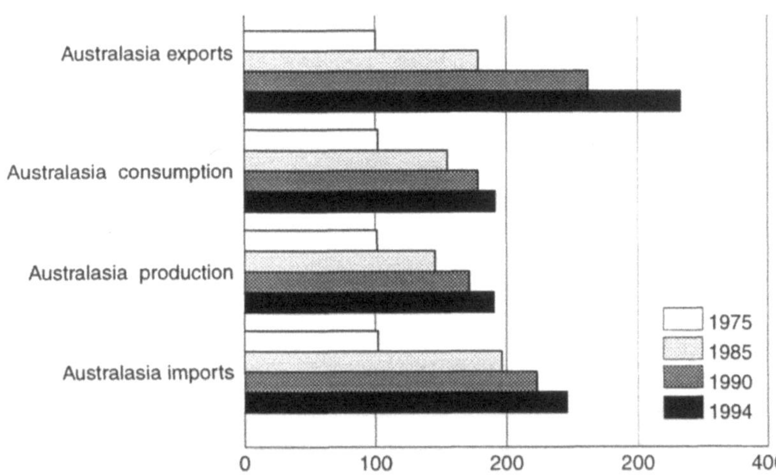

4.9 Australasian paper and board export, consumption, production and import indexed to 100 in 1975.

The indexed graphs show steady but not exceptional growth between 1975 and 1994 of around 4.8% per annum with percentage growth in exports just overtaking imports by 1994. Still the absolute import tonnage is considerably in excess of the export tonnage.

Asia

The picture in Asia is quite different. With a total consumption in 1994 (Fig. 4.10) of 79 million tonnes - almost 30% of the world total - and with a production of 71 million tonnes Asia had a net import of 8 million tonnes. In fact 12 million tonnes were imported and 4 million tonnes were exported.

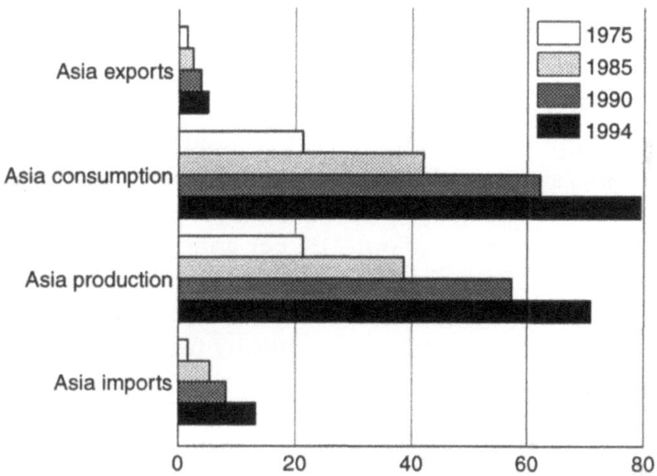

4.10 Asian paper and board export, consumption, production and import (million tonnes).

It is interesting to list the principal players in this big region. Their 1994 production and consumption figures rank them as in Table 4.1. Quite clearly many of these countries, for example Hong Kong (now part of China), Israel and Singapore, have no paper and board industry of any size. In fact only Japan and China produce substantial amounts of paper. The other countries doubtless make a viable product but cannot produce the variety of grades that they require and that the major producing countries have. Consequently they are heavily reliant on the import trade. This must be seen by other regions as providing an excellent opportunity to import their grades into Asia.

Table 4.1 Principal 1994 import, production, consumption and export figures across Asia (million tonnes)

Country	Import	Production	Consumption	Export
China	3.2	21.4	24.3	0.3
Hong Kong	1.1	0.2	1.3	–
India	0.3	2.3	2.6	–
Indonesia	0.2	3.0	2.4	0.8
Iran	0.5	0.3	0.9	–
Israel	0.4	0.2	0.7	–
Japan	1.2	28.5	28.8	0.9
Korea	0.6	6.5	6.1	1.0
Singapore	0.8	0.1	0.6	0.3
Taiwan	1.2	4.2	4.7	0.7
Thailand	0.6	1.7	2.1	0.2
Turkey	0.3	1.1	1.3	0.1
Other	1.7	1.8	3.5	–
Total Asia	**12.5**	**71.3**	**79.3**	**4.5**

Figure 4.11 shows the steady increase of production and consumption in Asia of around 12.9% to 14.7% per annum from 1975 to 1994. Not surprisingly, since production did not keep up with consumption, imports showed an average growth over the period of around 51% per annum and exports grew at about 21% per annum.

Consideration of the figures suggests that the growth rate has been high and that these growth rates will continue to increase. This represents a significant business opportunity for many years to come.

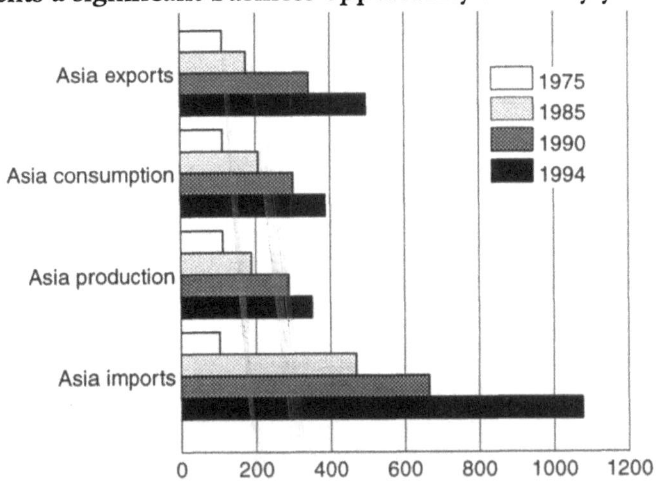

4.11 Asian paper and board export, consumption, production and import indexed to 100 in 1975.

North America

In 1994, North America produced 37% or 99 million tonnes of the world's supply of paper - and consumed 34% or 92 million tonnes of this supply (Fig. 4.12 and 4.13). It was a net exporter, exporting 29% of the world's exports or 21 million tonnes, while importing just 14 million tonnes. There is a major trade between the United States and Canada.

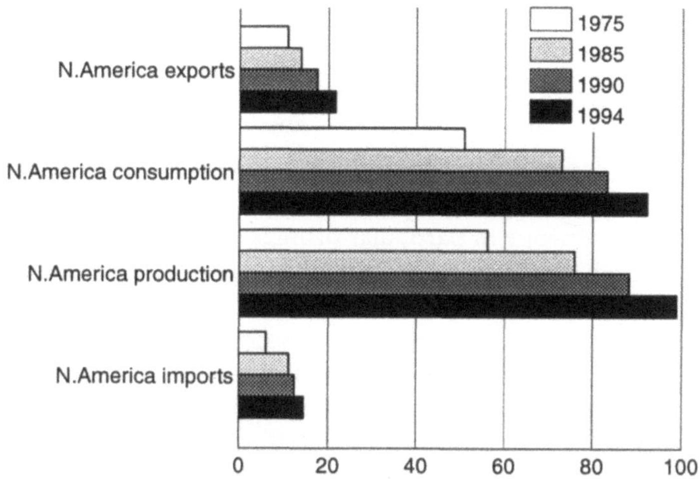

4.12 North American paper and board export, consumption, production and import (million tonnes).

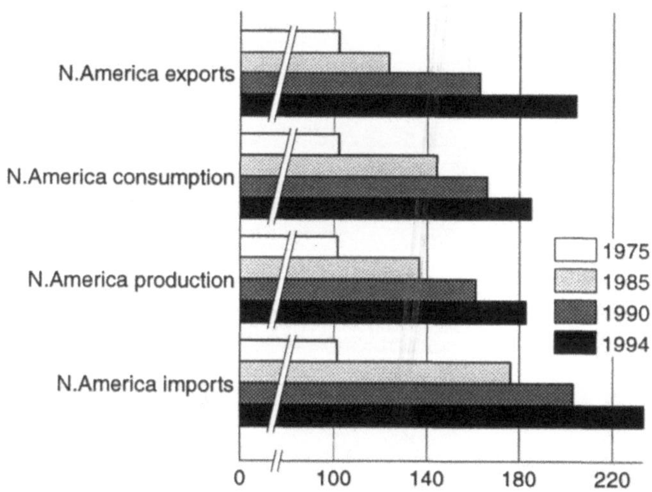

4.13 North American paper and board export, consumption, production and import indexed to 100 in 1975.

The tonnage that underpins these figures is very large, and the production and consumption growth rates on average over the period 1975–1994 have been about 4.1%. The statistics (Fig. 4.13) show a mature and stable trade. It is unlikely that the per capita consumption will increase greatly, compared with other countries. Nevertheless the tonnage is so great that even small growth will represent substantial opportunities for the mature paper industry in North America.

Europe

The European scene is quite complicated, and the position is even more difficult following the changes in Europe in the past decade which offered Eastern Europe in particular promises of better living standards. In its turn this uncovers the very large problems of technological updating of its manufacturing industries, and other improvements that are required in the paper industry of Eastern Europe.

First of all, let us consider Europe as a whole. Like North America there is a large manufacturing industry - North American production in 1994 was 99 million tonnes. The total European production in 1994 was 81 million tonnes, some 30% of the world total. North America comprises two countries but many states. Europe comprises 36 countries with different national characters and administrations. Figures 4.14 and 4.15 shows the production, consumption, import

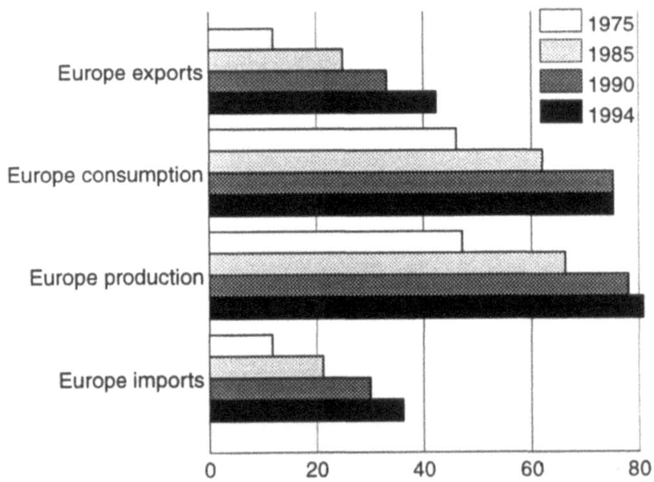

4.14 European paper and board export, consumption, production and import (million tonnes).

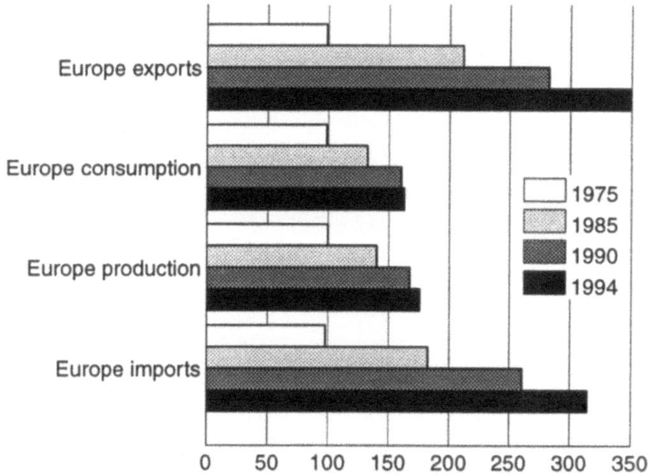

4.15 European paper and board export, consumption, production and import indexed to 100 in 1975.

and export statistics. The impact of the European recession shows in these figures, since between 1990 and 1994 there was barely any increase in production or consumption.

Europe is a net exporter outside its region, exporting 41 million tonnes and importing 36 million tonnes in 1994, giving a net export balance of just 5 million tonnes. This is, however, rather a simplistic summary as the import and export statistics for individual countries within Europe vary considerably. In particular the Scandinavian figures are noteworthy, importing 1.1 million tonnes and exporting 19.7 million tonnes, showing that they have substantial net exports of around 18.6 million tonnes, from a 22.4 million tonne production and 3.8 million tonne consumption.

The average growth per annum taken from 1975 to 1994 was only 3.3% in consumption, 3.8% in production, while imports were 11% and exports 13%. Within Europe, cross-border import and export has clearly been an important factor to this substantial papermaking community.

The split between Eastern and Western Europe statistics is shown in Fig. 4.16 and 4.17. The import/export trade in Eastern Europe barely exists, and production and consumption are substantially less than 20 million tonnes, and are on a falling trend.

While growth in Western Europe is around 6%, Eastern Europe shows no growth, and the figures in fact show a shrinkage. It will be interesting to see what changes will be seen in these figures in the next decade or two. The potential for considerable growth is there, but the time scale is very uncertain in view of other priorities on the limited resources of the Eastern European countries.

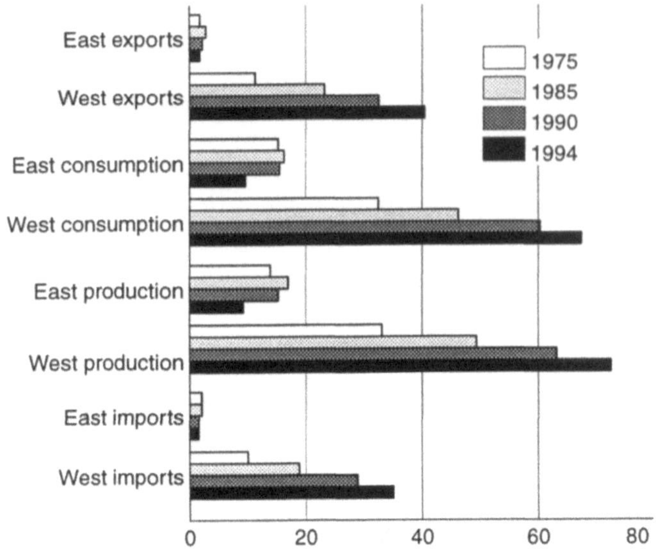

4.16 Eastern and Western European paper export, consumption, production and import (million tonnes).

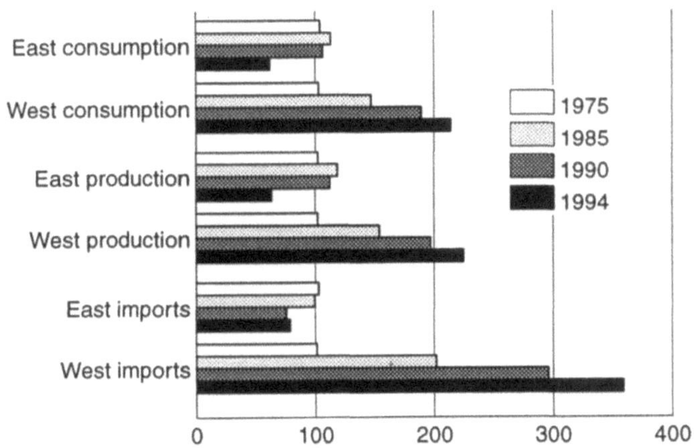

4.17 Eastern and Western Europe paper consumption, production, and import indexed to 100 in 1975.

All regions

The statistics of activity within the papermaking regions considered are best summarised for 1994 by Table 4.2, which shows the percentage of the world production, import, consumption and export.

Asia remains the most interesting region to watch since consumption and production may be expected to grow, while imports could also be expected to rise. Europe is the second most interesting region, taking as it does one half of the world's imports and exports. The situation will become even more important when the Eastern European situation is successfully unravelled, and the consumption and production in Eastern Europe rise. Other political changes in Europe will also have their impact, particularly the common market and common currency developments. Finally, North America remains the biggest consumer and producer and will continue to support a large import and export trade. Even though the growth in North America will not be the greatest in the world, the base level of operation will remain notably high. A start to realising the potential in Africa is unlikely to be made in the next ten years. Latin America, on the other hand, will continue to show more growth.

Table 4.2 Percentage of world share of import, production, consumption and export in each region

	Import	Production	Consumption	Export
World	100	100	100	100
Africa	2.4	0.9	1.3	0.8
Latin America	5.7	4.4	5.1	3.0
Australasia	1.8	1.2	1.4	0.9
Asia	17.9	26.5	29.6	6.3
North America	20.4	36.9	34.4	30.1
Europe	51.8	30.1	28.2	58.9

Summary

The import and export trade of 70 million tonnes of paper and 30 million tonnes of pulp in pulp and paper was worth roughly US$100 billion in 1994. Asia, and particularly China, are again highlighted as the region and country with the greatest immediate growth prospects, followed closely by Europe, and Eastern Europe in particular, although the time scale of this growth is more uncertain.

CHAPTER

5

Forestry

The parameters of forestry and the links between forest area and yield to paper production are discussed, giving an indication of the area of forest required to produce pulp for the annual consumption of paper, and the anticipated incremental area. International timber resources are discussed with particular examples drawn from Russia, Scandinavia and Canada, and the uses and management of the forest resource is considered. The chapter then focuses on pulpwood, its source and supply. The question 'Is there enough forest?' is considered.

Introduction

The world uses 173 million tonnes of pulp to produce 276 million tonnes of paper. The Northern Hemisphere provides 83% of this world's pulp supply. The forests therefore require and have a high and extremely important profile in the pulp and paper industry.

Forests are headlines. Forestry is a politically sensitive issue. Consider the following statement:

> *Despite an unprecedented ruling by Russia's Supreme Court, logging still goes on in the eastern part of Siberia's vast taiga. ... a joint Russian – South Korean operation ripping out virgin forest at what seemed to be an unstoppable rate.*

This news feature appeared under the heading 'Logging Destruction' in a recent issue of the *Geographical Magazine*.

People have built up a perception of forests that incorporates many of the features of the following statement which are more often than not both political and emotive.

> *The forests are being destroyed, and all of the users of wood, like the pulp and paper industry, are held responsible for their mass destruction. The rain forests are being destroyed, and cutting, clearing and burning are all responsible for environmental disaster. This disaster manifests itself through soil erosion, land slip, flooding in the wet and crippling drought in the dry, and the release of carbon dioxide to the atmosphere, causing the greenhouse effect.*

These are the negative points of the argument but there are also many advantages. This chapter sets out to put some shape and quantification into the subject of the world's forestry. It discusses the forests and particularly the wood supply in relation to, among other users, the pulp and paper industry.

Timber resources

Parameters of forest measurements

For the layman to understand the global forest problems and opportunities available to the forestry industry, it is necessary first of all to understand the terminology, and the systems of measurement and counting, and to relate them to the various factors under consideration, and in this case to the paper industry. Forest measurement is in unfamiliar units, and it is worth taking a few moments to discuss them.

Forests take up a considerable area of the world's surface, and produce a tree yield per hectare. Forests grow trees that are measured by volume, while the paper trade has an output measured in tonnes. The linkages between the following parameters need to be defined:

- Forest area.
- Tree yield per area.
- Tree yield - volume.
- Paper tonnage.

Forest area

The forest area is measured in hectares (ha), where:

- 1 ha is 100 ares or 2.5 acres.
- 1 ha covers 10 000 square metres (m^2).
- 1 km^2 is 1 000 000 m^2 or 100 ha.
- 1 ha is 1/100 km^2.

Tree yield per ha

One hectare will typically yield 1.5 m^3 of wood from a natural forest. The figure from a plantation will be several times, perhaps as much as five to ten times, larger than this. Tree yield is measured in cubic metres and combining the yield from both forest and plantation, an average yield of 3.6 m^3 of wood per hectare does not seem unreasonable.

Paper quantity

This is measured in tonnes. By considering the density of the wood fibre from which pulp is made, and the process by which it is made, a yield of pulp may be defined. For example, by the mechanical pulping process, 2.5 m^3 of wood are required to produce a tonne of pulp, whereas if the pulping process is chemical, the requirement increases to 4.5 m^3. Also, different wood densities affect the conversion, but on average, it takes about 4 m^3 of wood to make a tonne of wood pulp.

Table 5.1 starts from pulp, and shows the amount of wood in cubic metres required to make a tonne of pulp and the area of forest that this occupies. The final column shows the amount of paper made from that pulp. The statistics for the 1995 pulp consumption of 173 million tonnes is similarly treated, and the last row of the table shows the incremental pulp consumption, assuming a 3% increase of consumption per year.

Table 5.1 shows that the paper consumption of 276 million tonnes in 1995 required about 700 million m^3 of wood for its manufacture, which in its turn would require a forest area of 200 million ha or 500 million acres in which to grow the wood under present conditions of forestry. If growth of paper, and therefore pulp supply, is 3% per annum, then allowing for the present level for the reuse of 38% recycled fibre, the incremental forest requirement (Table 5.1) is 5.8 million ha per annum or 60 000 km^2 per annum.

Table 5.1 Area of forest, and volume of wood required for pulp and paper production

Pulp	Wood	Forest area			Paper
tonnes	m^3	ha	acres	km^2	tonnes
1	4.1	1.12	2.9	0.012	1.6
173 million	708 million	194 million	501 million	2 million	276 million
5.2(3%) million	21 million	5.8 million	15 million	0.06 million	8.3(3%) million

Table 5.2 considers the world forest area, compared with the total world area. 'Forest to pulp' means the amount of forest taken up for pulp production in 1995, and this is 5% of the available forest, and 2% of the available land. If one then considers the increments required by a 3% increase in paper consumption, one can calculate the 'forest increment' which is 0.06 million km^2, or a further 0.2% of the forest area.

Table 5.2 Forest area statistics (1989–90)

	Area (million km^2)	Forest area (%)	Land area (%)
World land area	130	–	100
World forest area	39	100	33
Forest to pulp in 1995	2	5	2
Forest increment for pulp	0.06	0.2	0.05

The figures in this model are very simplistic, and may be used only as crude indicators. They are included to show the magnitude of the subject being dealt with. But now, having got some perspective into the statistics, it is appropriate to consider a more detailed analysis of forest resources.

Forest resources

The dominance of Asia (23%) and North America (USA and Canada) (22%) in the world's forest resources can be readily appreciated (Fig. 5.1) particularly with by far the greatest forest concentration occurring in North America, while the African continent possesses some 15% of the world's forest resource, and former USSR (FUSSR)

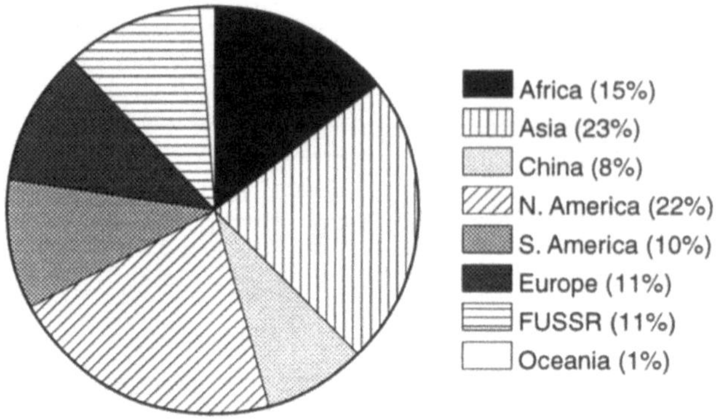

5.1 The relative sizes of the world's forests (%). (Source: FAO Forest Products Annual Year book 1993, FAO, Rome, Italy.)

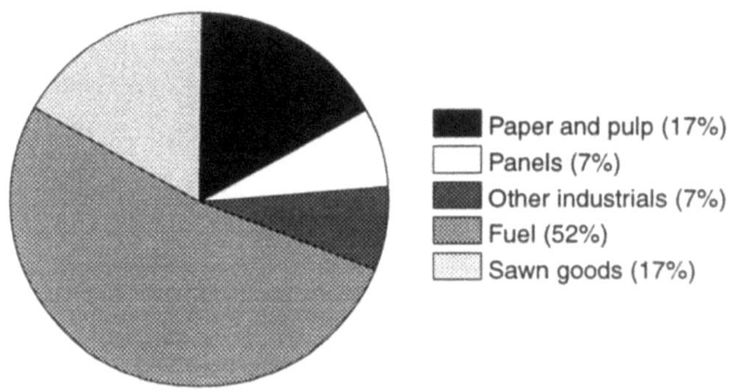

5.2 Wood consumption across the world (%). (Source: FAO Forest Products Annual Yearbook 1988, FAO, Rome, Italy.)

some 11%. The contribution of Australia and New Zealand, the main Oceanic countries, is just 1%.

The destruction of the forest is highlighted by pressure groups, and is frequently drawn to the public's attention. It also causes the pulp and paper industry much concern. Figure 5.2 shows the enormous amount of forest resource, 52%, burnt as fuel. The problem that is caused by this amount of incineration is compounded by the release of carbon dioxide into the atmosphere from the carbon that was locked in the wood. Forest is also lost through clearing the ground to allow the extensive grazing of livestock in, for example, the meat-producing areas of South America. This also shows in the statistics as the use for fuel, but the real cause is clearance for grazing.

The world needs good forests. They bind the land together, preventing erosion and land slip. They absorb carbon dioxide from the atmosphere, and it should be noted that young trees are more effective at this than mature trees. They provide a sump for carbon,

Table 5.3 Total wood consumption projected to 2010 for all uses in developed and developing countries, billion m^3 (Source: FAO 1995)

Year	Developed countries	Developing countries	Total consumption
1993	1.25	2.09	3.33
2010	1.80	2.86	4.66

Recycled content of paper 38% (1993) projected to increase to 45% (2010).

which prevents it from being present in the atmosphere as carbon dioxide. They provide industrial benefit, through effective logging operations, and trees from forests are important as a fuel material. Besides this, they are a recreational amenity, and through their friendly environment they provide medicines and foodstuffs - honey, mushrooms, fruits and berries, clover and hides.

It has already been seen (Fig. 3.3) that by the year 2025 the figure for the world's population is expected to exceed 8 billion. The population is growing fastest in the East: in Asia, 60% of the population is less than 25 years old. This will put increasing and considerable strain on the forest resources, as they are used by this new population (Table 5.3). The population is also becoming more literate. In China literacy is now 80%; in India it is 50%, and a consequence of greater literacy is a greater expectation for an improved standard of living.

Taking these factors together, they indicate a scenario of more people, and a greater per capita consumption of paper, leading to a greater strain on the forest resources.

The 2010 total consumption is derived from the generally anticipated growth at 2.3% per annum (Nilsson S 1996). This is an average position, but there is in fact a wide variation in growth statistics, country by country, so it is quite difficult to use an average figure with any great meaning. To give precise figures would demand more detailed research, and often the basic information is just not available.

Forests need good management, and this is a priority for the effective control of the resource. But in order to make it come about, resources must be provided so that management of the forest can take place. Once they are in place, economic reward can be taken from the invested effort and material. To quote UK Foreign Office development plans, 'A forest without industry is of no financial value to government, but the social and environmental and ecological values are enormous'. As a model, it is worth noting the way in which the Nordic,

North American and Australasian forests are managed. In each of these cases, forest management is economy-driven. In each case there is a significant industry relying upon the forest, and long-term planning requires attention given to the raw material resource, so that the industry will continue. In these countries forests have been managed with some success for some decades by the pulp and paper trade.

When considering efficient management, the priorities for action are in the following forest uses:

- Forestry in land use - management, agriculture, forestry and more rational land use.
- Forest-based industrial development.
- Fuel wood and energy.
- Conservation and ecosystems.
- Institutions - strengthen public forest administration and professional and vocational training of forestry extension and research.

Efficient management is more likely to come about if an economic case has been presented than if we have to rely upon philanthropy. Efficient management can then be set to achieve the following:

- The provision of fuel wood for 2000 million people.
- The promising prospect for agricultural production through conservation and the rational use of forest land.
- The expanded processing for employment and income.
- The increased production for energy.
- Conservation.

There is interesting documented information in the selected areas of Russia, Scandinavia and Canada.

Russia

Comment has already been made on the increasing amounts of FUSSR timber that will become available to the year 2020. In round terms Dr Aslund, Professor and Director of the Stockholm Institute of Soviet and East European Economics, quantifies the timber availability from the East of Europe in 1993 as:

- Soviet timber: 360 million cubic metres.
- Swedish Timber: 60 million cubic metres.
- Polish Timber: 20 million cubic metres.

As a measure of significance the forest area in Poland occupies about one-third of the total land area of Poland, and the forest is comparable

Table 5.4 1990 per capita consumption

	Paper and board (kg per capita)
Sweden	239
FUSSR	36
Poland	24–36

in size to the Swedish forests, so that they have the same potential. Aslund says that 'In the long run, Poland should be one of the big sources of timber in Europe'.

In Russia, wood transportation is uneconomic, the average timber travelling distance, from point of harvesting to point of use, being 1650 km. Harvesting is badly uneconomic, and 50% of the wood is wasted. It is lost at the forest, which also contains large areas of rotting timber. Not only this, but Soviet paper mills require five times more timber per tonne of paper than Swedish mills, principally because of the continued use of old technology. Only 33% of the wood cut in Russia gets to the stage of being shipped so that Russia requires 15 times more timber per tonne of paper than does either Finland or Sweden.

The Ukraine currently harvests 10 million m^3 a year, and Belorussia produces 7 million m^3 a year. Russia now exports most timber to the Far East, Japan and North Korea. It also exports 4 - 5 million m^3 to Finland, 1 million m^3 to Hungary and 1 million m^3 to Austria.

The FUSSR, the Czech Republic, Slovakia and Poland will be able to export more timber in the future. It is worth comparing the per capita consumption of paper and board in Sweden, the FUSSR and Poland, to see also the potential for greater consumption in the latter two countries. The figures available are for 1990 and are shown in Table 5.4.

One may draw the conclusion from this that there are large timber resources in the FUSSR and the Eastern European countries, and that were the forests to be better managed, and the technology of papermaking brought up to date, there would be ample supplies of timber, even though paper consumption may greatly increase.

Scandinavia and Finland

Liisa Saarenmaa's article 'The ecology of boreal forests' (1995) is also pertinent. Scandinavian forests are mainly in the Boreal zone, and conifers dominate. The total land area is 102 million ha and 50% is forested. Species of Scots pine and Norway spruce were the original conifers in Scandinavia, and white and silver birch grow well. Finland has some 45 000 species in all.

Forest fires make a great impact on the boreal forests, and they occur with a frequency of every 50 years or so on dry sites and every 120 years or so on moist sites. All forests in Finland have been burnt at least once in the last 400–500 years. Some species may only seed and grow after a serious fire has taken place. Gap disturbances are being studied. It is said that clear felling mimics forest fires. In that case it is necessary to leave some decaying wood in areas of clear felling for certain insects to inhabit.

Sustainability is an evolving concept and, particularly, sustainability occurs where the total removal from the forest system does not exceed the yield. But today, sustainable forestry is not enough - it must also be ecologically sustainable.

Canada

Canada would like to be seen as an eco-friendly supplier of market pulp, and it sets out the following conditions under which it operates its forestry development programme. Sustainable forestry involves managing to achieve the following objectives:

- Protection of the forest ecosystem.
- Maintenance of forests of differing ages and species.
- Protection of fish, water and wildlife.
- Maintenance of recreational values.
- Maintenance of economically viable forest industry.

A little over one-third of Canada's forests will be used as a wood source. Foresters calculate the annual allowable harvest - rather like the controversial European fish quotas - and companies are not permitted to exceed them. In calculating the quota, items such as age, growth rate and environmental considerations are taken into account, and they are regularly reviewed. Currently the Canadian harvest exceeds 169 million m^3, which is 75% of the total allowable. However, the total figure allowable will be affected annually by fire, epidemic and disease.

Industry is committed to the concept of a renewable resource. Growth is greater than the rate of disaster and harvest. Half of the forest regeneration is natural, and the remainder is from seedlings grown from indigenous species.

The area of clear-cutting sites is controlled. The areas vary depending on local circumstances from 5 to 120 hectares. In this way, the areas concerned are designed to mimic the results of natural disasters, but again another school of thought is developing as in the Nordic regions, that clear-cutting over even these areas should be avoided.

In the past ten years 17.4 million ha have been regenerated and this rate has ensured that the amount regenerated exceeded the amount harvested by 8.1 million ha. This is good, but large amounts are not by themselves sufficient. It is the biodiversity of the forest that must be preserved. Canada is also active in promoting sustainable forestry world-wide.

In concluding this section it is perhaps worth observing that the challenge of the future must be to bring the Eastern European and the tropical hardwood forests to the same level of managed efficiency as the better forest areas of the Northern Hemisphere. In order to put more pressure on the requirement for 'good management', the World Wide Fund for Nature (WWF) and the Forest Stewardship Council (FSC) recommend and are pushing for the certification of all forests, to qualify them for an acceptable provenance.

It is interesting to note that the IIED survey of 1993 found that the paper cycle is a net contributor to greenhouse gas emissions. While plantations store large amounts of carbon on land that was not previously forested, there is not yet sufficient to offset emissions from fossil fuel used in manufacture, and methane emissions from paper that goes to landfill. The contribution is about 1.7% of total global carbon dioxide emissions and equivalent. This is in comparison to steel and chemical industries contributing 4.6% and 5.9% respectively. The paper industry deficit could be offset by, for example, planting 76 million hectares of plantations, without allowing for improvements from other sources.

Pulpwood from the forest

Roundwood is a term used to describe all wood in the rough, without taking into account whether it is for fuel or for industrial use. It is all wood felled or harvested from the forest and the trees outside the forest.

A number of researchers have made estimates of the world's

Table 5.5 Roundwood across the world, and conversion to paper

Country	Roundwood (billion m^3)	17% to paper (billion m^3)	Paper tonnes (1990) (million tonnes)
World	3.5	0.60	238
Africa	0.5	0.09	3
Asia	1.1	0.19	56
North and Central America	0.8	0.14	91
South America	0.3	0.05	8
Europe	0.4	0.07	67
Oceania	0.1	0.02	
Other	0.3	0.04	13

roundwood supply, and information from the FAO shows the total availability in 1990 of roundwood (Table 5.5). The figures are obtained by assuming that across the world, 17% of roundwood is used by the paper trade and finds its way into paper. Another worker, Dr Nilsson (1996), finds that of the world forests, 64% are coniferous, and 36% are non-coniferous. He gives the distribution of the total wood supply by region and the projection to 2020 (Fig. 5.3).

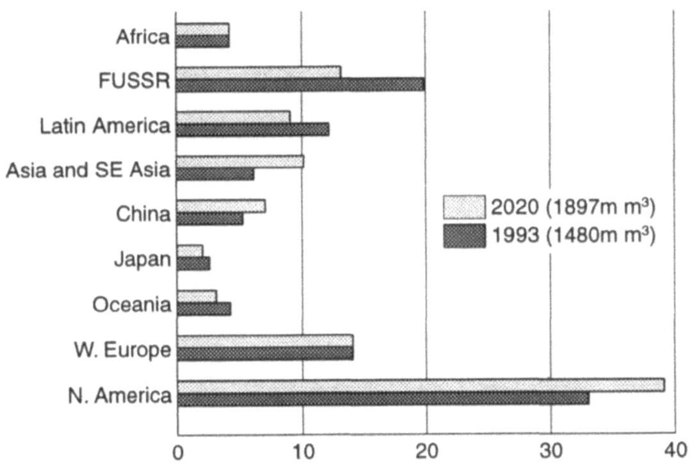

5.3 Total wood supply (% by region).

The indications that may be drawn from this can again only be tentative. Overall the wood supply increases, but the statistics point to a reduction of the total percentage coming from Asia, and particularly China, and in North America. This is more than offset by strong increases from the FUSSR and Latin America. The various sources of information point at least in part, to the need for improved forest management and utilisation in Asia, and in the FUSSR where some of the improvement is showing in the statistics. Although the percentage from North America is falling, it remains a powerful wood producer. As far as pulp production is concerned, the international paper trade will remain strong in North America, and Western Europe, while Latin America and the FUSSR will produce substantially more than hitherto.

Is there enough pulpwood?

Pulpwood is not seen as prime timber, and it has its place in the wood hierarchy after veneers, sawnwood and panels. It takes up around 17% of the roundwood supply (Table 5.5). The locations from which it is collected are shown in Fig. 5.3, which clearly shows the effect of a pulp industry in forest development. The USA and Canada are big in pulp and papermaking, as are Scandinavia and the FUSSR. Each has well-developed pulping and papermaking industries, and they provide the wood pulp for over 75% of the world's demands.

Wood for paper

Whether our forest resources are sufficient is difficult to assess. One must consider the forests from which roundwood is derived and pulp made, and the paper that is produced from the pulp. The distribution of the world's forests and of the roundwood supply is shown in Fig. 5.4 and 5.5, and Fig. 5.6 shows the distribution of the world's paper production.

From this information, some interesting deductions can be made. Roundwood, from which much pulp is made, is obtainable in quantity from FUSSR but is inefficiently exploited. The FUSSR has 11% of the world's forests, supplying 18% of the world's roundwood, and producing 4% of the world's paper. We have already said the management of the FUSSR's timber resource leaves much to be

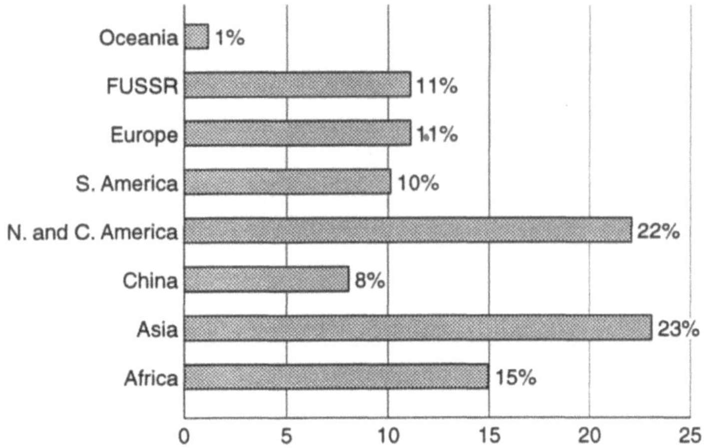

5.4 World forests (% by region) (Source: FAO Yearbook).

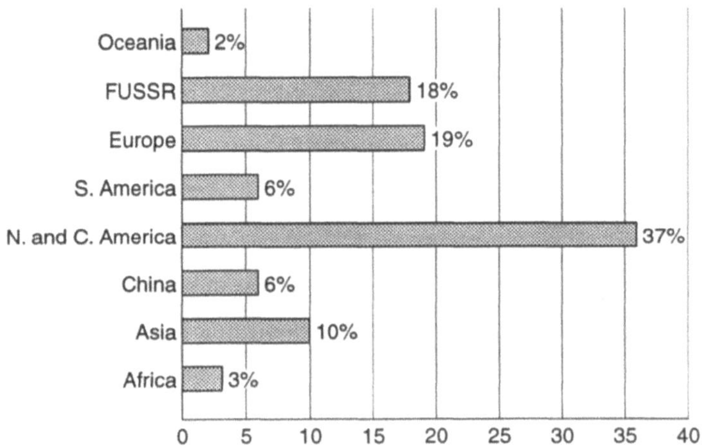

5.5 World roundwood supply (% by region).

desired, but improvements are taking place, and the FUSSR is again becoming an important player in the world forestry scene. Roundwood exploitation is better in Europe and in North and Central America. This resource must also be readily available in Africa and Asia (including China), and it should provide a good opportunity for its conversion into wood pulp.

Africa and Asia have forest resources that could be made to yield more if they were better managed.

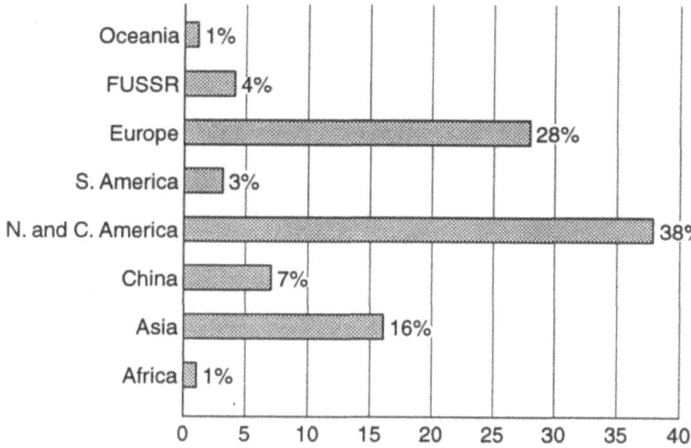

5.6 World paper supply (% by region).

Land utilisation

Land utilisation and forest management play under the same banner. It has been shown that better utilisation is achieved through better management. It is well recognised that the forests of North America, Western Europe, Latin America particularly Brazil, and Australasia are considerably better managed than those in the remainder of the world.

The forests of the FUSSR, Poland and other Central and Eastern European states are developing and could be improved, both politically and economically, and much effort is expected here in the middle to long term. They, together with Africa, Asia and parts of Latin America, need more and better management of their forest land.

There is every reason to believe that there are sufficient pulpwood sources, particularly with the promise of better forest management and plantation development giving better yields. This will in its turn give an improved environment, and the availability of fuel possibly by plantations grown and replaced specifically and efficiently for this purpose. With greater managed efficiency, forest land can be made available for agricultural and urban clearing, while doing no damage to the environment. These are targets that the pulp and paper trade should actively support in pursuing the aim of a better environmental profile, which must be of the greatest importance.

Summary

Possibly the best summary of this section is provided by the IIED survey of 1993, published in 1996 ('Towards a Sustainable Paper Cycle', IIED, 1996). It examined the sources of fibre because there was an absence of global information on this matter. It revealed the following:

- Managed natural regeneration forests are the single largest source of wood fibre (37%).
- Unmanaged natural regeneration forests account for 17% of wood fibre supply.
- Plantations provide 29% of global wood pulp.
- Original conifer forests account for 15% of total wood pulp.
- Tropical rainforests provide only 1% of global wood pulp and original hardwood forest elsewhere in the world provides another 1%.

The industry believes that its environmental problems are in control. The forests are a managed resource - they have to be. Their management involves regeneration. What still needs discussing is the environmental and economic role that plantations could in the future and in some cases already do fulfil.

CHAPTER

6

Raw materials

Numerous raw materials are used by the pulp and paper industry and this chapter provides an overview of the principal ones. The aim is to give some idea of quantity and usage of raw materials in the industry, and perhaps to highlight the trading opportunities, without providing a definitive list. Cellulose fibre is essential to the paper sheet. Different types of fibre, for example hardwood and soft-wood, eucalyptus, esparto and straw, rag and so on, have different properties. The broad approach is to discuss fibrous raw materials which provide the basic structure of the paper sheet, followed by minerals, which are perhaps the second most important raw material classification after cellulose, then water, the use of which is of paramount importance to the industry. After this the variety of chemicals used is discussed.

Fibrous raw materials

In discussing the different varieties of fibre it is necessary to consider the total world fibre utilisation, as the statistic shows an interesting, though not unexpected, trend over the past 20 years. The increase in world paper consumption figures has been noted a number of times, and of course the fibre utilisation figure must show a similar increase. If one constructs a materials balance, there are a number of components that contribute to the weight of a tonne of paper. First of all there are the major components – virgin fibre, often

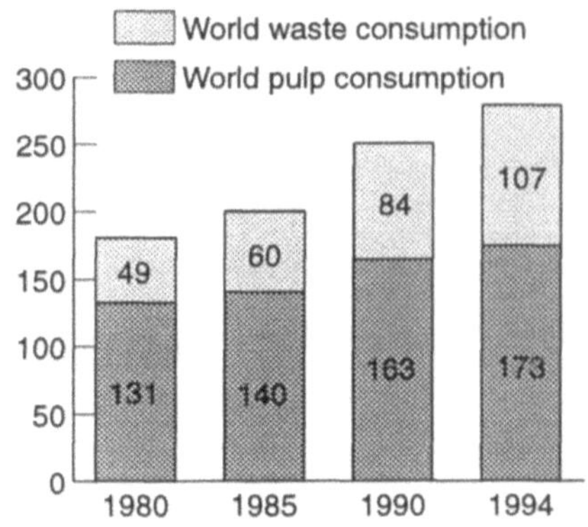

6.1 Waste paper and pulp consumption worldwide (million tonnes).

from wood pulp, recycled fibre and mineral content. Besides this there is water, contained in the fibre, or introduced in the process, and also remaining in the finished piece of paper. Then fibre and mineral are lost during the papermaking process. The amount remaining is different for different fibres, minerals and grades of paper, but it is usual to consider a 'retention' figure of solids that remain in the paper. All of the factors involved make the exact fibre and mineral retention figures difficult to calculate, but typical fibre retention figures are 85 - 95% with mineral retention much poorer than this.

Figure 6.1 shows the increasing amount of fibre used per annum since 1980. Consumption of paper has increased and so the amount of fibre used has increased. In 1994, 276 million tonnes of paper were consumed, and 173 million tonnes of virgin fibre and 107 million tonnes of recycled fibre were used to make it, together with 20 million tonnes of mineral, giving an average solids retention figure of 92%. The increase of fibre consumption was greater for recycled fibre than for virgin pulp.

Figure 6.2 expresses these figures as an index based on the 1980 value as 100%. Up to 1985 the virgin fibre component was increasing more rapidly than the recycled fibre component, but since 1985 the rate of use of recycled fibre has outstripped that of virgin fibre. Since 1980, total waste paper usage has increased by an average of 8.5% per annum, while average virgin fibre usage has increased by about 2.7% per annum.

As will be seen in the next chapter, waste utilisation in different grades ranges from nothing at all, as for example in photographic papers, where it would be detrimental to the production of a blemish-free photographic image, to a utilisation of 100%, in grades such as newsprint and corrugating medium. Calculating a global figure for

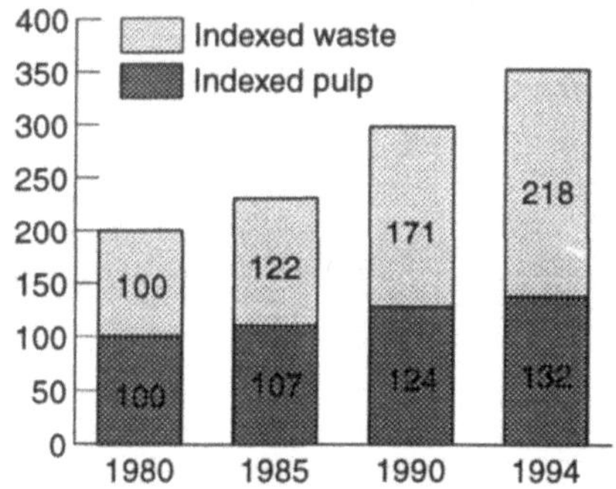

6.2 World pulp and waste paper utilisation indexed to 100 in 1980.

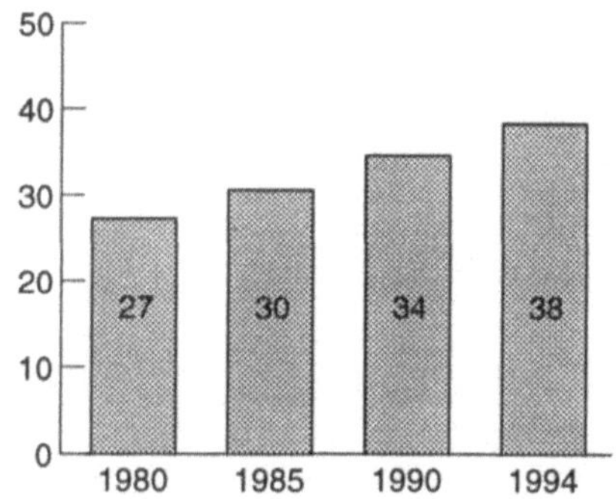

6.3 World recycled fibre utilisation (%).

recycling, Fig. 6.3 shows that by 1994, recycled fibre usage in paper manufacture had increased to around 38% from around 27% in 1980.

When the paper statistics were discussed earlier, the world was divided into regions. Moving now to pulp, the split is much simpler. The principal geographical regions remain the same, but within those regions there are substantially fewer manufacturers of pulp than there are of paper. In particular, pulp production is located more in the Northern Hemisphere than in the Southern. Figure 6.4 shows the geographical split.

In absolute terms, the major producing regions in Europe are the Nordic Countries and the Commonwealth of Independent States (CIS), with several other countries contributing. In Asia the major producers are China and Japan, while in Latin America, Brazil is the biggest producer.

It is perhaps surprising to note in Fig. 6.5 how few countries are significant pulp producers and this of course is a reflection of their possession and management of forests. The USA is by far the biggest pulp producer, producing with Canada almost 50% of the world

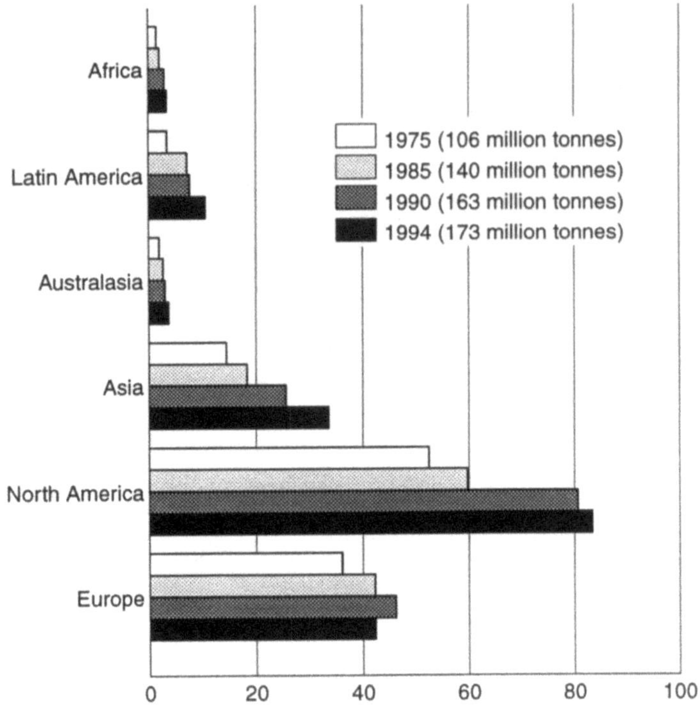

6.4 World-wide pulp production by region (million tonnes).

supply. The third supplier is the Nordic countries with 13.4% followed by Japan with 6.1%, Brazil with 3.5% and the CIS with 2.6%. The last two may be expected to increase their pulp production considerably in the next decade. Of the producers classified as 'other', over half are to be found in Europe.

So far little distinction has been drawn between the different varieties of pulp. Of the total pulp production of 173 million tonnes in 1995 only 10 million tonnes was produced from non-wood fibres.

Wood pulp

By far the greatest amount of pulp (around 95%) is based on cellulose fibres from wood. The grades are split into mechanical, chemical and semichemical grades and their classifications are largely self-explanatory.

Mechanical grades have their fibres separated by the mechanical means of grinding or refining. The pulp contains many impurities, and the lignin content is prominent. The lignin makes the pulps hard to bleach and their colour is not permanent. That is why an old piece of

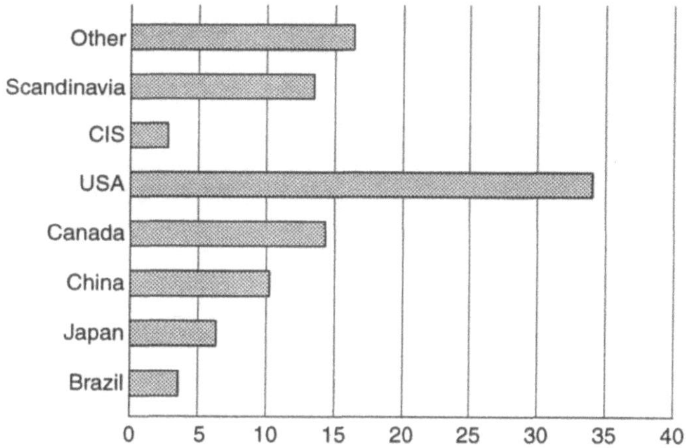

6.5 World pulp production in 1994 (%).

newspaper is yellow. The discoloration is caused by the presence of lignin which yellows on exposure to sunlight.

There are principally five varieties of mechanical pulp. They are:

1 Stone groundwood, or groundwood pulp (SGWP).
2 Refiner mechanical pulp (RMP).
3 Thermo mechanical pulp (TMP).
4 Chemimechanical pulp (CMP).
5 Chemithermomechanical pulp (CTMP).

The *SGWP* process has been established since 1840 and is still widely used. Debarked softwood logs are pressed against a rough rotating stone grinder. The fibres are torn from the logs, and some oversized pieces are generated which are washed, refined and returned to the main fibre stream. No lignin is removed and many resins and extractives remain in the pulp. The yield is 95% and the paper produced is bulky and opaque. A further development of this process is to surround the grinder with a pressurised case and to operate the grinder at higher temperature. This variation is known as pressurised groundwood.

The *RMP* process was developed in the 1960s. Softwood chips are passed through refiners, between two grooved iron discs which are coaxially grooved and which rotate in different directions. They operate in one or more stages. The fibres receive less damage than in stone groundwood and so are longer, while retaining the same yield.

The *TMP* process is a further development where the wood chips are preheated in steam to 110–150 °C for up to an hour before refining takes place.

The *CMP* process combines the above process with chemical

impregnation of the chip using sodium sulphite but at a lower temperature of around 30–40 °C. The process may be used with hardwoods. The yield is in the 85–90% range, and a much stronger pulp than RMP may be produced.

The *CTMP* process uses a light impregnation of sodium sulphite followed by thermomechanical chipping. The process, first introduced in 1978, produces a yield of 90%. The pulp produced is a little lower in bulk and opacity, but the strength is higher.

Paper strength is promoted by the contact and bonding of hemicellulose in the fibres. Good contact usually means a strong bond. When pulps are prepared mechanically, the fibres separated may be damaged or reduced in length. The fibres will also be in contact with the lignin since it has not been removed, and the fibres are stiff. These factors impede good fibre-to-fibre contact and so the opportunity for good hemicellulose contact between the fibres is interfered with. Thus mechanical pulps give rise to paper that is not strong and is of poor performance and colour. They are therefore cheaper pulps with a selling price depending very much on the strength of the market, but around US$300 per tonne.

Two types of chemical pulps are produced. They are produced by the sulphite process and the kraft or sulphate process. They do not have the deficiencies outlined above. Chemical treatment ensures the lignin is removed in the cooking liquor - for example caustic soda, or sodium sulphide - and the fibres are separated chemically. This ensures that the fibres are not damaged, that they are more flexible and that they can develop strength through good bonding. The strongest pulps are produced by the kraft process, and they are simply classified as kraft pulps. Pulps in this group normally sell at up to US$600 per tonne, although there is a wide variation with figures of US$500 to US$1000 being noted over the past few years.

The *sulphite process* is a traditional acid sulphite digestion process. There has been a great deal of development of cooking methods in recent years, but it is not really relevant to review this work here. Fibre yields fall in the range 40–55%, and resultant cooking and bleaching effluent has become more environmentally acceptable by the selection and development of improved chemical treatments. Nevertheless lack of convenient and economic chemical recovery leaves environmental pressures on this operation, although paradoxically the resultant pulp is brighter and can be made high white without extremes of bleaching conditions, which is an environmental advantage.

The *kraft process* is dominant in chemical processing. The basic process applies to both hardwoods and softwoods. The main chemicals used are sodium sulphide and sodium hydroxide, with salt cake (sodium sulphate) being used to make up sulphur and sodium loss. It is the use of sodium sulphate that gives the process its name. A softwood kraft retains more lignin (3.8%) than a hardwood kraft (1.3%), thus reducing the bleaching load in the production of a bleached kraft.

Eucalyptus pulp deserves a special mention. The fibre is produced on fast-growing trees and its pulping was developed in Australasia. The fibres are short and not very flexible so that the papers produced are normally bulky and not very strong. Eucalyptus forests are now developing across the world and there are centres of eucalyptus pulping, not only in Australia and New Zealand, but also in countries where the trees grow fast, notably in Brazil, Spain and Portugal.

Semichemical pulping is again, as the name implies, a hybrid between mechanical and chemical pulping. In semichemical pulping, some mechanical treatment takes place after milder chemical treatment.

Bleaching

Most unbleached pulps are brown. However, there are quite wide differences in colour, depending on the source of the fibre, the pulping process and the efficiency of washing to remove often deeply coloured cooking or grinding liquors.

The cellulose and hemicellulose in wood is predominantly white, but the lignins and resins are coloured and need to be removed in the bleaching process. Lignin, modified by either chlorination or oxidation, is dissolved by sodium hydroxide. With oxygen delignification, it is practicable to recover sodium as carbonate, and some heat by reintroducing the liquor into the kraft recovery cycle. Chlorination, and even chlorine dioxide use, does not have this advantage, and the extracts from chlorination pose a considerable pollution load for waste treatment systems.

Principal bleaching agents, used singly or in combinations, are:

- Chlorine.
- Oxygen.
- Hypochlorites - both sodium and calcium.
- Chlorine dioxide.
- Sodium dithionite (hydrosulphite).
- Ozone.

With environmental concern about chlorolignins in effluent, and discharged to rivers, there is a strong tendency to move away from the use of chlorine. Indeed the total consumption of chlorine has been halved since 1988, despite the considerable growth in pulp production since that time. Other work being carried out is the treatment of sludge to reduce chlorine emissions.

To summarise, the wood species, pulping method and bleaching method all give the pulp different characteristics, and the papermaker has a wealth of pulps available to enable him or her to select the ones that suit the manufacture of different varieties of paper.

Non-wood raw materials

The utilisation of non-wood raw materials represents only about 5% of the world's pulp supply, and so they tend to be used only in special situations. The world capacity for non-wood pulping is twice this amount. Theoretically, all cellulose-containing materials can be used to manufacture paper, but the more likely ones are esparto grass, cereal straw, bagasse, rice straw and cotton linters. There are then other fibres, usually for speciality grades, such as hemp, rag, abacca, sisal and kenaf fibres. Estimates from a United Nations Food and Agriculture Organization (FAO) Capacity Survey show the following capacities for 1993 (TAPPI Journal **79**(10) 1996):

- Straw - 9.6 million tonnes.
- Bagasse - 3.0 million tonnes.
- Bamboo - 1.3 million tonnes.
- Others - 6.9 million tonnes.
- Total non-wood capacity - 20.8 million tonnes.
- % non-wood capacity - 10.6%.

The problem with the majority of non-wood fibres is that they are bulky, and they need to be gathered over a short period of time, yet have to be kept and used for a year while the next crop is growing, although the position is better than this in some locations, where there are three crops each year. Viable pulping almost always implies high-volume pulping and that, together with the fact that non-wood pulping usually has a poor yield - something like 35% - and the raw material is available only at harvest time, lead to problems with the logistics of supply. The straw pulp has an unwanted silica content and besides this the effluent containing by-products of cooking and bleaching needs careful treatment before disposal.

The main cooking process is also based on the kraft process as used with wood fibres and caustic soda is the principal cooking chemical. Bleaching also follows the wood pulp bleaching technology.

Table 6.1 Straw pulping capacity

Country	Straw pulping capacity (million tonnes)	World total (%)
China	8.4	88.0
India	0.28	3.0
Pakistan	0.26	2.7
Spain	0.12	1.3

About 100 years ago esparto grass used to be a major papermaking fibre, but it is an expensive fibre today and is only manufactured in North Africa and Spain. In the 1910s and later in the 1940s European esparto production turned to the use of straw, and this proved to be a versatile papermaking fibre. The raw material has its champions today, but it has still to establish itself as a major raw material, and its popularity as a potential fibrous raw material fluctuates. Nevertheless, environmentalists subscribe to the idea of non-wood fibres as raw materials, and straw is the most popular material. The most important countries for straw pulping capacity taken from an FAO Capacity Survey in 1993 are given in Table 6.1. Of the other countries examined, they all had substantially less than 1% of the world total capacity.

In the speciality field, there are some good examples of non-wood fibre utilisation. For example, cigarette tissue is manufactured from hemp and some rice straws, and some high-quality writing and artist's papers may be made from cotton linters. But taking both wood and non-wood cellulose fibres into consideration, the quantities involved in straw utilisation, including the single high Chinese figure, or other non-wood grades are not yet highly significant.

Minerals

Minerals are used in the industry for a number of reasons. They are available in large quantities in certain parts of the world. They give an economic advantage since a tonne of mineral used as a filler costs an average of around US$200 while a tonne of fibre costs on average US$600. There has therefore been an economic desire to increase the percentage of mineral loading in the paper. Although increased loading in the paper causes a reduction in the strength of the paper, it improves its printability by making the paper surface smoother, and the surface can be left matt or given a gloss by calendering. Minerals are also used extensively in coating formulations for the same reasons since again they can be used to enhance the gloss and printability of some grades.

Historically, the most commonly used mineral is china clay (kaolin). It is used both as a filler and as a coating material. Calcium carbonate is cheaper and is gaining ground since it can be used where neutral sizing is used. Precipitated calcium carbonate is a special filler, produced at the paper mill, using waste carbon dioxide and slaked lime, and talc is available from France and Finland. Titanium dioxide is used as an opacifier in small portions in many papers, or as the main filler in certain speciality grades. Coloured pigments may also be used in certain speciality grades.

1 *China clay (kaolin).* China clay, hydrated aluminium silicate, is a versatile mineral, and varieties suitable for use in paper are found in Brazil, Germany, the UK and the USA. It is inert, and comes from the feldspar in granite deposits. The primary material is found in the UK, in Cornwall, where it is washed out of the deposits where it is formed. Jets of water are used to wash out the clay and the material is then processed by settling and screening, before using chemical and centrifugal cleaning to form the filler material. A secondary material is found in the USA and Brazil. Its inertness is advantageous in that it does not react with other materials in the paper or coating furnish in which it is used, and it can be manufactured in a variety of particle sizes and brightness for its different applications. Some exploitation of kaolin deposits found in Australia is now also taking place.
2 *Calcium carbonate.* This is white and cheap, and there are abundant sources. It is restricted to use in alkaline suspensions, since acid systems react with calcium carbonate, but with the advent of neutral sizing techniques its use is becoming more widespread. It occurs naturally, but it can also be produced in a precipitated form where it has a special use particularly in cigarette tissue manufacture. Used in large quantities of over 20% retained in the sheet, it will give opacity to the cigarette tissue paper, and it also imparts desirable control of burn rate characteristics to the cigarette.
3 *Talc.* Some talc is used as a filler, but it is oleophilic, and is therefore not suitable for use in paper which is used subsequently for offset printing.
4 *Titanium dioxide.* This is a high refractive index, inert pigment, conferring on paper a very high degree of opacity. It is an

expensive pigment, and is used in some commercial printing papers as an opacifier in small quantities of about 1% of the total furnish, in combination with clay or chalk. It is also used in large quantities in the manufacture of white plastic base papers, where the titanium dioxide content can be as high as 13–26%. It is available in two forms, anatase and rutile. It is more readily available in the USA, and it is cheaper there, so that it is used in the USA in greater quantities.

5 *Coloured pigments*. Inorganic and organic pigments such as barium sulphate (barytes) and zinc sulphide are used to make opaque brightly coloured papers for the manufacture of coloured laminates, or for special applications such as the manufacture of baryta-coated as opposed to resin-coated photographic paper.

6 *Coating pigments*. China clay and calcium carbonate are both used in the manufacture of coated papers. The mineral forms an inert opacifying layer of excellent printability and can be used to produce either a matt or a gloss finish. Typical coat weights are 10–12 g/m^2, and the mixes or 'slips' as they are called, contain the pigments mixed with other chemicals, including binders to assist the uniform application of the coating slip.

Some 21 million tonnes of mineral were used in the manufacture of 276 million tonnes paper in 1995. Table 6.2 shows the quantities and percentages of the main raw materials used. The table has been constructed taking into account the fact that different materials have different moisture contents. Paper is in equilibrium with the atmosphere, where it contains about 7.6% moisture content. Pulp and recycled fibre (RCF) are usually quoted 'air dry' or containing about 10% moisture. The table shows that 278 million bone dry tonnes of raw material are used to make 255 million bone dry tonnes of paper. From this it can be seen that the retention of solids is around 92%, or

Table 6.2 Principal raw material content of paper (1995)

Raw material	Quantity (million tonnes)	Percentage	Bone dry (million tonnes)	Water (million tonnes)
Paper	276	100	255	21
Pulp	173	57	157	16
Recycled fibre	107	36	96	11
Mineral	21	7	25	
Total solids	**301**		**278**	

in other words, it requires 1.09 tonnes of bone dry raw material to make 1 bone dry tonne of paper.

This illustrates the concept of retention, described earlier, where not all of the material fed to the paper machine appears in the paper, and some of it is lost with the water that is extracted from the paper at the draining section of the paper machine. Reuse of this material has in the main been incomplete and a typical mill retention figure is 90–92%.

Another technological point is worth mentioning. It will be seen that there are about a hundred chemicals used in the manufacture of paper and so there are quite complex charges on the ions in suspension or solution. Modern technology is providing an understanding of the chemical charges and equilibria of these systems so that application of this technology can ensure that more of the materials added actually remain in the paper.

These factors lead to some development projects that are working to improve retention with the objective of operating with no effluent. The projects are known as zero effluent projects. Success in this work is both environmentally and economically desirable.

Water

First of all, let us tabulate the water usage parameters as they may be applied to papermaking, so defining the average amount of water needed to make a tonne of paper, and the total amounts used by a large paper machine, as a total volume and as a rate of use per minute. Besides describing the total amount of water being handled, consideration is given to a more usual situation of 90% recycling of water, leading to only 10% use of freshwater in the system.

Process water

It has been emphasised earlier in this book that water is essential to pulp and papermaking. Cellulose fibres form a slurry with water and they are transported around by pumping them through pipes. Pressing fibres together in water promotes bonding and strength development between fibres, and once the water has been used to transport the fibres it becomes redundant to the cycle of the operation and is either reused or rejected. Water is also necessary for steam

Table 6.3 The parameters of water

Weight	Volume
4.5 kg (10 lbs)	1 gallon
1 tonne	1 m^3
	220 gallons

Table 6.4 The parameters of water usage in paper production

Paper	Total water use	Freshwater water use (90% recycled)
1 tonne	300 m^3	30 m^3
	300 tonnes	30 tonnes
	66 000 gallons	6600 gallons
250 000 tonnes	75 million m^3	7.5 million m^3
0.5 tonnes/min	150 m^3/min	15 m^3/min

raising as the paper once produced has a proportion of its drying carried out on steam-heated cylinders. The quality of both the process and the steam-raising water needs to be carefully controlled by suitable chemical addition. The parameters of water are given in Table 6.3.

It can be seen from Table 6.4 that, allowing for 90% recycling of process water, a rate of freshwater usage of the order of 15 m^3/min (3300 gallons per minute) is necessary for a mill manufacturing 250 000 tonnes per annum. This rate of usage requires a total supply of 7.5 million m^3 of freshwater a year. It is essential therefore to start with a reliable, clean, well-organised water supply to the mill and a good effluent treatment plant at the end of the mill, be it a pulp mill, a paper mill or an integrated mill.

Even with a suitable supply, water treatment is essential to control pH, dissolved matter, turbidity and colour in the water. This control is carried out in settling beds, and by chemical treatment certainly with alum and possibly other chemicals to flocculate impurities and control pH before passing the water through final settling beds. Consideration must also be given to water hardness and the need to add chemicals to soften the water. It may be necessary to treat the process water generally, but it will almost certainly be necessary to treat the boiler water to prevent damage to the boiler.

Most paper mills run systems that are closed, certainly to some degree. They may be closed to a high degree and it has already been noted that the 'zero mill effluent' system is now being researched. One of the problems is that a closed system causes a general build up of anionic trash, and slime problems can develop which will cause paper breaks. A programme of slime treatment will be required in order to have an essential break-free operation of the paper machine. Nevertheless make-up water will always be required and so water treatment will continue to be necessary.

Effluent treatment

Water that has been used cannot just be dumped to drain. It requires treating before it can be returned to the stream, sewer or system from which it was extracted. Treatment again depends upon how polluted the water has become but it is usual, after as much of the solids and heat as possible have been recycled, to settle, sediment and clarify the effluent, ensuring that the water returned is of acceptable quality, and settling the suspended solids so that they may be dealt with by themselves. Extracted solids may then be sent to landfill, or possibly to incineration. All systems will need to conform to discharge within rigorous standards of suspended solids, biochemical oxygen demand (BOD), chemical oxygen demand (COD) and absorbable halogen (AOX).

The commonly used terms in relation to effluent are:

- Active chlorine - chlorine that contributes to the bleaching effect.
- AOX - absorbable halogen.
- BOD - biochemical oxygen demand. The power of biological and other components in an effluent to remove dissolved oxygen from the water.
- COD - chemical oxygen demand. The power of the chemicals in an effluent to remove dissolved oxygen from the water.
- EOX - solvent-extractable organic chlorine.
- TOCl - total organic chlorine.
- TOX - total organic halogen.

Chemicals

In Chapter 1, the value of the total amount of chemicals and minerals consumed in the pulp and paper industry was identified as US$19 000

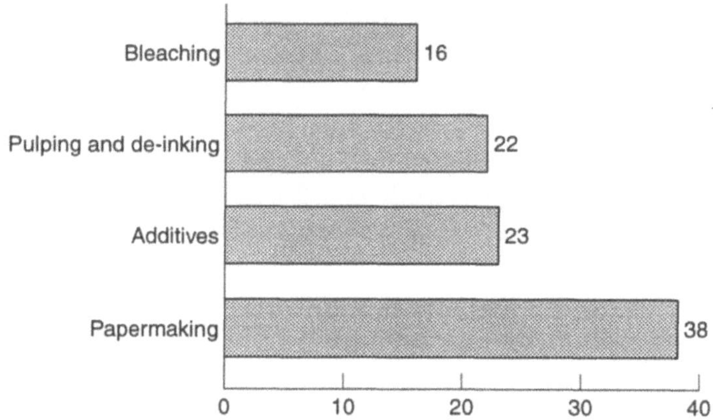

6.6 General pulp and paper chemicals by value for the USA in 1993 (% of US$4700 million).

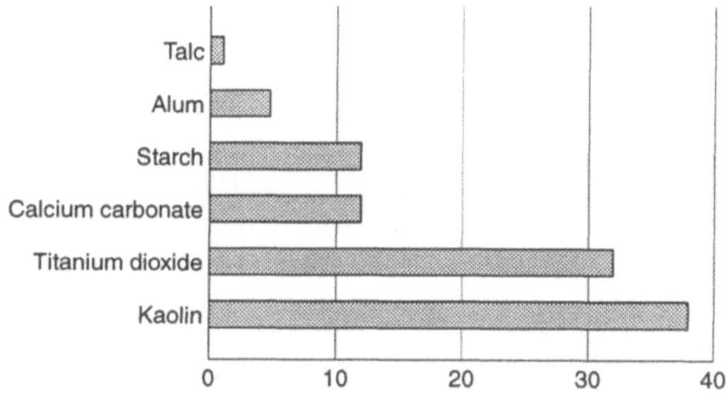

6.7 Paper making chemicals by value for the USA in 1993 (% of US$1800 million).

million, with chemicals amounting to around US$14 000 million. The breakdown of those figures into categories is interesting. The USA accounts for roughly 40% of the world production, and the breakdown of US consumption by value is now considered.

Figure 6.6 shows the general breakdown in the USA of the consumption of chemicals in 1993. Papermaking is the most important consuming process, using almost 38%, while additives at 23% and pulp and paper de-inking at 22% are followed in economic importance by bleaching which represents 16%, at a total value of about US$4 700 million.

Figure 6.7 considers this diverse group in a little more detail. The breakdown for papermaking chemicals, which include chemicals for basic paper production, such as kaolin, clay, starch, calcium carbonate, titanium dioxide and talc, is shown. The products mainly act as fillers and replacements of the more expensive fibrous components, but they are also very important in increasing brightness and opacity.

Minerals have already been dealt with but their importance in paper production as fillers and coating materials is underlined quantitatively. Alum and starch are also widely used. However, the importance of alum is starting to decline as more producers switch to alkaline papermaking and neutral sizing. Despite this decline, alum remains a significant material in water treatment, in pitch control, in wet strengthening and as a colour fixer.

Starch is an important papermaking additive, and its use is increasing. This is caused in part by the increased use of lower-quality waste, where it is used as a wet end additive to improve the performance of the product. It adds strength and improves retention, and by applying it as a surface layer at the size press, it improves surface strength.

The next important classification, accounting for US$1100 million in 1993 is the group of pulp and paper additives (Fig. 6.8). These are speciality chemicals used to impart special sheet properties to the paper, or to enhance existing values. They include special sizing agents to control the penetration of liquids into the paper, wet strength additives particularly for sanitary grades of tissue, dyes and pigments to impart colour, and processing chemicals such as defoamers.

The most valuable group is dyes and pigments, accounting for 24% of the additive's total value, although the environmental considerations are moving some of the market away from bright

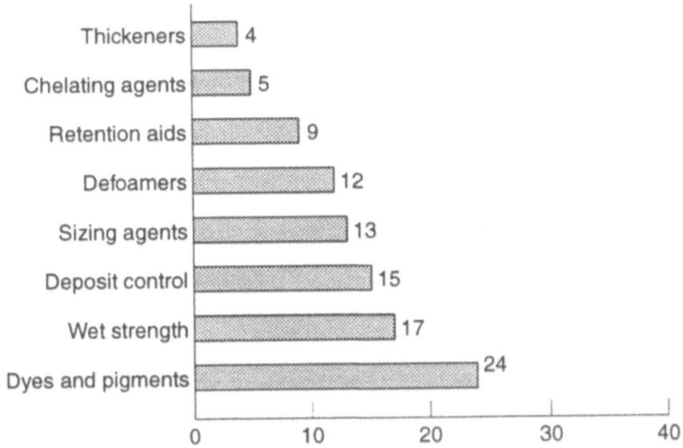

6.8 Pulp and paper additives by value for the USA in 1993 (% of US$1100 million).

colours. Optical brighteners - chemicals absorbing ultraviolet radiation and converting it to additional visible blue and white light, thereby enhancing brightness - are also important in the dyes and pigments group.

Wet strength resins are of continuing importance (17% of the total value) and much work is in progress on their environmental compatibility. They are frequently used to enhance paper strength, based on the ability of the resin to retain the tensile strength of paper even after the complete saturation of the product with water, and some are also used to enhance the dry strength.

Both deposit and foam control are important to the papermaker, and together this group was worth, in the 1993 US paper market, 27% of the total value of US$1100 million. Deposit control agents are of increasing importance where the industry is moving to more recycling and zero effluent systems, which will increase the amount of contaminants in the process. Typical contaminants are slime, metallic deposits, adhesives and stickies (contaminants other than adhesives that adhere to papermachine clothing and interfere with production and machine efficiency). The use of defoamers falls into a similar category of usage, and they are widely used across the pulping and papermaking processes to improve the efficiency of these processes through trouble-free operation.

Sizing agents (13%) are important, particularly with the requirements of high-quality papers. Sizing agents limit the rate of water penetration, enhance burst and tensile strength, and reduce surface picking on printing. They also play their part in helping to provide high print quality.

The third detailed chemical category is provided by chemicals used in pulping and de-inking (Fig. 6.9). The group is dominated by caustic soda which accounted for 76% of the US consumption of US$1000 million in 1993. Caustic soda is used primarily as a cooking chemical, removing lignin from wood, mainly in chemical wood pulp production. Clearly, with the emphasis on the use of recycled fibre, the use of de-inking chemicals to assist in the removal of printing inks from the paper surface, is of great, and growing, importance.

The last classification in this group is the use of bleaching chemicals. These have been influenced by the environmental changes of recent years, leading to the greater use of peroxide and sodium chlorate in place of chlorine, which is considered environmentally unfriendly. The statistics for the USA showed usage of bleaching materials worth US$750 million, and the breakdown is shown in Figure 6.10.

While still playing an important part as a bleaching chemical,

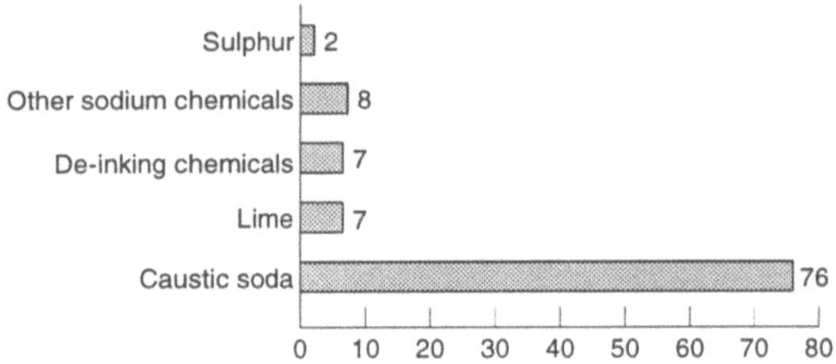

6.9 Pulping and de-inking chemicals by value for the USA in 1993 (% of US$1000 million).

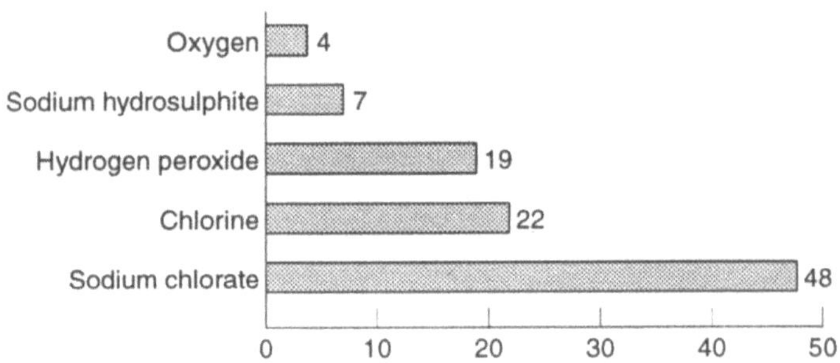

6.10 Bleaching chemicals by value for the USA in 1993 (% of US$750 million).

chlorine has lost its place as the most important bleaching material. It now represents 22% of the share of the US$750 million bleaching chemicals market in the USA, and totally chlorine-free pulps are now becoming popular in Europe. The switch away from chlorine is expected to continue.

Sodium chlorate at 48% has taken much of the market occupied by chlorine, and it yields a brighter pulp, while creating significantly less fibre degradation and effluent emissions. However, it is a chlorine-derived substance, and its environmental effect could have a detrimental effect on its usage.

Oxygen and hydrogen peroxide are good competitive bleaching agents. Of the total value of all chemicals used by the industry in the USA, 89% by value was used in paper and board mills, and 11% was used in pulp mills.

Some of the more commonly used chemicals are listed in Table 6.5, but there is an ever-increasing number.

The use of chemicals in the paper industry is an essential part of the process, and they give rise to an increase in businesses based on the paper industry. The materials are important because they allow the

Table 6.5 Chemicals commonly used in the paper industry

Pulping	Cooking	Sodium hydroxide, sodium sulphide, calcium bisulphite, magnesium bisulphite, sodium carbonate, sodium sulphite
	Bleaching	Chlorine, oxygen, calcium or sodium hypochlorite, sodium chlorite, chlorine dioxide, hydrogen peroxide, sodium sulphite, calcium and magnesium bisulphites, zinc and sodium hydrosulphites
Papermaking	Retention	Polyamides, polyacrylamide, polyethylene imine
	Colour	Dyes and pigments
	Optical bleaching agents	Fluorescent whitening agents
	Sizing	Rosin size, aluminium sulphate (alum), alkyl ketene dimers (AKD), starch, cationic starch
	pH control	Alum, sulphamic acid, sodium hydroxide
	Pitch dispersion	Alum with talc or bentonites, anionic dispersing agents
	Wet strength	Urea formaldehyde and melamine formaldehyde, modified starches
	Slime	Organosulphur compounds
	Defoaming	Emulsified mineral oils, self-emulsifying fatty acid derivatives
	Cleaning	Grease absorbers
Coating	Adhesives	Binders, polyvinyl alcohol (PVA), casein, glue, starch, CMC (carboxy methyl cellulose)
	Latexes	Latex-based products usually in coating mixes
	Water	Alum flocculants, activated silica
Oiling and greasing		Standard lubricants and greases

development of enhanced paper properties, and they are also responsible for making available the technology to keep fibre and mineral in the paper, and to reduce the load on effluent treatment plants. There are also now a number of chemical suppliers who are substantially dependent on the pulp and paper industry, as the examples in Table 6.6 show.

Pulp and papermakers are also substantial customers of the oil

Table 6.6 Chemical suppliers' dependence on the pulp and paper industry

Chemical	Main Use	Business in paper (%)
Chlorate	Bleaching	90
Latex	Coating	53
Kaolin	Filler and coating	50
Carbonate	Filler and Coating	40
Peroxide	Bleaching	40

Source: CJC Marketing, Norwich, 1996.

industry in the use of their products for the lubrication of complex machinery that is required to operate continuously and with little attention throughout the year.

Biotechnology

Biotechnology exploits the use of biological organisms, such as enzymes and similar materials, in industrial processes. Enzymes are defined as proteins that act as catalysts in biological reactions. There are around 3000 of them, and they are very specific in the conditions under which they catalyse reactions. They are so far relatively uncommercialised in the pulp and paper industry, but are well known in the fermentation industries - for example, in cheese, beer, wine and bread making.

They would seem to hold promise to be the elixir of the industry, but it must be said that they do have an impressive record of achievement. They are not yet used widely in the pulp and paper industry, partly because, as has already been commented upon, the industry is so highly capital intensive and production orientated that there is understandably an unwillingness to take any risks that could jeopardise the high productivity called for. In spite of this, there are already proved cases where the following improvements caused by enzymes can be substantiated:

- Higher productivity.
- Lower operating costs.
- Limitation and reduction of waste.
- Reduced cost of waste treatment.

A number of applications have been suggested where biotechnology may improve the efficiency of pulp and paper manufacture. Biotechnology, resulting in genetic engineering can be introduced into the forest, resulting in tree engineering, and indeed, is being introduced. Because trees have a long growth period to maturity - some 40 to 60 years being typical, these changes can only be slow. Nevertheless the faster growing Eucalyptus forests (8 to 10 years to maturity) will be the first to see the beginnings of change.

There are already developments where very specific problems are being successfully addressed, such as the resistance of certain strains to herbicides, and the improvement of rooting characteristics. Genetic engineering of trees has been mentioned earlier, and this work is being carried out in Eucalyptus research where there are large plantation crops, so that the work could have considerable impact in the next couple of decades.

The impressive list of benefits claimed from this work and extensions of it are:

- Reduced energy costs per tonne of pulp.
- Reduced chemical consumption for cooking and bleaching (for example by making the lignin content easier to treat).
- Increased cellulose yield per tonne of pulp.
- Improved fibre quality and mill throughput.
- Reduced effluent.

Biotechnology can then work in the pulp mill where enzymes may for example be sprayed on to wood chips preventing staining of the wood, and so make debarking easier, resulting in the saving of energy, and controlling pitch and stickies.

Enzymes also assist in bleaching where they reduce the amount of bleaching chemicals used, and in some cases assist in producing high levels of brightness without the use of chlorine. Currently there are examples of this from around twelve mills worldwide that are in some form of continuous production.

Stock preparation is another target for enzyme utilisation, and biobeating reduces the energy demand of pulp treatment, and where enzymes can be used for faster drainage of paper stock on the machine, faster machine speeds can be achieved. They are also used in waste paper recycling, by assisting de-inking and again saving chemical costs.

Energy

The pulp and paper industry is a large consumer of energy. Pulp mills use more energy per tonne than paper mills, but much of this energy is used again as heat energy. It is the greatest energy user in Canada, and the second greatest user in the USA. As an industry for consuming energy, it ranks fifth in the world. Energy utilisation has been referred to throughout the discussions on manufacturing. The proximity of pulp mills to hydroelectric schemes is indicative of this high consumption, particularly in debarking, chipping and refining. Paper processing is a direct energy consumer in, for example, refining and in the various items of water removal and through the drive of the paper machine and other items of plant, where even the smallest of paper machines will have over 100 motors.

Drying requires the generation of steam, and high-pressure steam has been used traditionally to drive an electricity-generating turbine, so that the pass out steam can be utilised for drying. It is interesting to note that in the early 1900s it was usual for this to be the case. By the 1960s only two-thirds of the mills in the UK followed this procedure and by 1990 the number had fallen to as low as 20%.

At this stage the most expensive bought in costs were for materials, followed by energy and manpower. Considerable attention has been, and continues to be, given to the problem of energy efficiency, and in the period of the last ten years, energy consumption has been reduced dramatically in the UK, with a consequent energy cost saving of 13% over 10 years, figures that are no doubt mirrored across the world.

Combined Heat and Power (CHP) projects are now assuming an important role again. Their demise had been partly because of the high cost of plant replacement and also because the grid had become more efficient. Their revival follows environmental pressure for high power production efficiencies and low emissions. The latest developments show the wider use of CHP plants. These are reputed to operate at 85% efficiency, compared with around 35% for a conventional power station, and typically they use around a third less fuel than conventional methods of electricity and steam generation.

Statistics published by the Paper Federation of Great Britain (PT 37(5)) show wide variation in energy consumption statistics from one manufactured grade to another. Examples of the range of energy consumption shows variation from 36 therms per tonne for waste paper conversion, to 202 therms per tonne for pulping and from 136 therms per tonne for the manufacture of corrugated case materials, to 439 therms per tonne for the manufacture of some speciality papers.

The industry has always been aware of the importance of controlling its energy costs. In the UK it has progressed from an average energy utilisation per tonne of product of 290 primary therms per net tonne in 1980 to a figure of 195 primary therms in 1994, a reduction of almost one-third.

The political and economic climate is again such that further reductions in total energy consumption will become a necessity in the next decade, driven by the lower targets being set for the further reduction of carbon dioxide emission.

Summary

The chapter has dealt with the raw materials used in pulp and papermaking. Recycled fibre in 1994 accounted for 38% of the fibrous raw material used, the figures being 173 million tonnes of pulp and 107 million tonnes of recycled fibre. Non-wood raw materials account for less than 5% utilisation, although there is a production capacity of twice this figure. China is the principal non-wood user, with a capacity for 8 million tonnes of straw pulp. However, the logistics and technical problems of straw production militate against its widespread adoption.

Different pulping methods have been briefly reviewed, and the predominance of North America and the Nordic countries noted as the major wood pulp suppliers. The 20 million tonne mineral utilisation was discussed and the different minerals reviewed.

Water into and out of the mill systems was discussed, and the chapter finished with a review of the principal process chemicals used in papermaking, in pulping and bleaching, and as additives, a total chemical business worth in the order of US$19 billion. It was noted that some suppliers are now heavily dependent on the pulp and paper industry.

Biotechnology applications are described and they appear to be poised to lead to improvements across the board in the areas of forestry, pulping and papermaking.

Energy is an important raw material, and energy reductions of almost one-third have been noted over the past 15 years or so. This reduction is expected to increase as more energy-saving measures are taken, driven by environmental targets.

CHAPTER

7

Recycling and recycling materials

The use of waste paper in the paper industry is discussed. Recycled fibre (RCF) is defined, and the chapter deals with the classification of waste paper and the content of waste in the various grades of paper. There follows an outline of the fibre recovery process, including de-inking, and the world-wide use of de-inked pulp. This is followed by a consideration of the 'urban forest' and the recovery of waste world-wide. The legislative pressures on recycling, and the environmental impact of these procedures is considered. The chapter concludes with some discussion of waste-paper pricing.

Introduction

World consumption of paper passed 250 million tonnes in 1993, and with an average growth rate of 3.1% per annum since 1975 it is on course to approach 300 million tonnes by the year 2000. In 1995, 276 million tonnes of paper used in its manufacture around 280 million tonnes of cellulose fibre, and around 20 million tonnes of minerals. This is the potential availability of materials to be recycled. Of course not quite all of this is available – for example, archival material, or paper that has been used in a manner to destroy its recyclability such as in the manufacture of resin-coated photographic papers, or silicone-coated self-adhesive backing sheets. Nevertheless, the maximum potential supplies are not far removed from this. Recycling is undertaken to recover both the fibres and the mineral content of the paper, while taking steps to remove unwanted material, such as printer's ink and adhesives and many other contraries (impurities).

It is inevitable that fibre is lost from the recycling potential. This could be because paper is treated so that it cannot be recycled, or it could be burnt or destroyed. Just because there are grades of paper that are made from 100% recycled raw material does not lead to the conclusion that there are raw materials available to allow 100% recycled content of all paper. Some people believe that only 70% raw material will be recycled, and this implies 30% of virgin material will need to be added. On subsequent stages of recycling the position is likely to deteriorate further and an even lower recycling percentage will be pertinent.

Paper and the use of waste

Paper is produced in a number of widely different varieties. Newsprint, boxboards, corrugated boards, printings and writings, tissues and speciality papers such as photographic paper and cigarette tissue, are just a few examples. Printed papers are often coated just to enhance print quality, the paper having a mineral-based surface coating applied at the paper machine, or in a separate process.

The variety of fibrous raw materials is wide, and its use may be specific to the grade of paper produced. For example, newsprint or boxboards require a less sophisticated raw material than do higher-quality printings and writings. The hardwood or softwood content, the original wood species, the pulp treatment and the bleaching treatment combine to produce grades that exhibit specific properties. There are also very special small quantity raw materials, for example rag, hemp, straw, all of which confer other special properties.

Recycling poses a particular problem, and with it an opportunity to the papermaker. For several decades some grades have been made from predominantly recycled raw material. Corrugating medium is a good example, and so also is board, where the internal plies of the board are from lower-quality recycled material. However, the market often requires special quality considerations peculiar to only some grades, for example, photographic papers cannot tolerate blemishes or contaminants in the product, which would interfere with the production of clear images, and certain food wrapping grades have a tight raw material specification precluding the use of recycled raw material in their manufacture.

Nevertheless the industry has always carried out a recycling programme, in addition to recycling its own waste generated and reused within the paper mill at which it was produced. True recycling levels are now increasing. Consider the fibrous raw material figures for the UK (source, the Paper Federation of Great Britain, Industry Facts, 1995). In 1984, of all fibre used, 43% was imported and 50% was waste paper pulp. In 1994 the imported figure had fallen to 31% while the waste paper figure had risen to 57%.

In 1995 the paper industry world-wide used 107 million tonnes of recycled fibre, which is 38% of the total fibre used. This was a great increase over the 1970 consumption, where only 30 million tonnes of recycled fibre was used. A combination of social, environmental, technical, economic and legislative factors have led to further substantial increases in RCF (recycled fibre) utilisation, and this will continue to be the case.

Current estimates suggest that within ten years, RCF amounts used world-wide will equal the amounts of virgin fibre used. The industry is responding to this by carrying out massive retrofitting operations to enable it to treat and use increasing supplies of waste paper.

The role of waste paper has gained respectability, moving from a liability as a waste, to a valuable resource, seen now as an asset. RCF pulp can now be made to a uniform quality, without contaminants, sweeping away many of the production obstacles to using the material in such a capital-intensive production industry that places big demands on high levels of efficient machinery utilisation.

Recycled fibre definitions

There have been many problems in defining the terms used in the waste-paper sector, not least being the use of the term 'recycled'. It is now recognised that it will be necessary to state how much of the fibrous content of the product is made from RCF in order to justify use of the term.

Her Majesty's Stationery Office (HMSO) in the UK has studied this problem in some depth, highlighting the wide range of political, commercial and other interests involved. These interests are represented in particular by:

- Extreme environmentalists calling for the abolition of virgin fibre from the supply chain.
- Papermaking opportunists offering 'recycled' paper made from predominantly own mill waste.
- Virgin fibre manufacturers denigrating the very use of alternative fibre.
- Recycled paper manufacturers defending market positions by insisting on hurdle rates as high as 75% recycled content to prevent market entry by new players.
- Government trying to reduce pressure on landfills (some 40% of domestic waste is paper) by convincing the public that recycling is environmentally sound.

To clarify the position, HMSO recently offered the following definitions of commonly used terms:

1 *Virgin fibre*: fibre derived from wood or such alternatives as bagasse and straw which are alternative fibres used to enhance paper properties. Cotton linters and sawmill waste should not be counted as recycled since they too are alternative fibre sources.
2 *Broke*: waste paper that occurs during paper manufacture. It has always been reused and this will continue. However, even if it has not been used by its originator, it should not be classified as recycled fibre.
3 *Pre-consumer/industrial waste*: any fibre source that may be counted as an industrial by-product. It encompasses such items as printer's trim, converter waste, reject materials and excess inventories. This fibre source is valuable and easy to collect in bulk. As a matter of good economics it should be directed back into the papermaking cycle. There is no desire to see this source artificially expanded to enable companies to satisfy the demand for recycled paper. With improved business practices under ISO 9000 and BS 7750 and with better planning, and economic design, and with shorter press run-up times, there is every possibility of minimising but not eliminating this source of waste fibre.
4 *Post-consumer waste*: fibre that has been used for its final and intended purpose. It is the largest source but the most difficult to collect. As virgin fibre prices rise, new techniques for segregating and economic processing of this waste are emerging.

Using the guidance provided by these definitions, HMSO recommend that to qualify as recycled, a paper must contain at least 10% post-consumer waste.

Waste classification

Almost any type of paper and board can be recycled, and comprehensive classification systems have been developed to ensure that maximum commercial value is obtained from recovered fibre. Classification systems are not internationally standardised, which make data comparison between countries difficult. The grading system produced by the Paper Federation of Great Britain is shown in Table 7.1.

Table 7.1 UK standard groups of waste paper

Group	Waste-paper type
1	White woodfree unprinted
2	White woodfree printed
3	White and lightly printed mechanical
4	Coloured woodfree
5	Heavily printed mechanical
6	Coloured krafts and manilas
7	New kraft lined, KLS
8	Container waste
9	Mixed papers
10	Coloured card
11	Contaminated grades

RCF differs from virgin fibres in a number of ways:

- In most cases waste paper consists of a mixture of different types of pulp fibre and other materials, such as mineral filler.
- Waste paper contains fibre that has normally undergone some type of mechanical refining during the primary papermaking process.
- RCF has been subjected to drying in the papermaking process, and possibly in conversion, for example heat set offset printing.
- RCF can have high contamination, in particular inks and stickies (adhesives, lacquer, etc.)

Recycling fibres back into the papermaking process implies reconditioning the material to its original virgin fibre condition as far as the

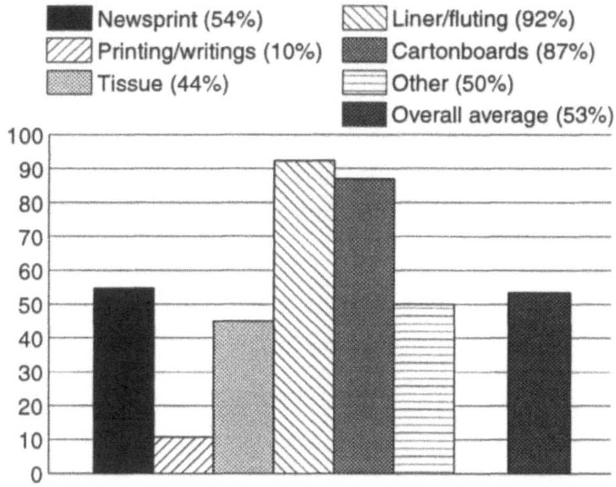

7.1 Current use of RCF in the range of paper grades in Europe.

reuse requires. However, some of the changes made to the fibre are irreversible, particularly those caused by refining and drying. A complete return of the fibre to its original state is not possible. Each time a fibre passes through the manufacturing process it suffers strength and colour reduction, although there is some evidence that the strength loss is just in the first few passes. Many people believe there will always be the need to add some virgin fibre to make up for the strength losses in the process, and to restore any loss of quality and runnability.

In general terms, recovery and reuse of RCF have concentrated on the easiest and most economical waste paper to collect and recycle, and this has been predominantly used in the lower-grade papers such as case packaging and board materials. Now that there is an increased demand for recycled fibre, there is a tendency to use RCF in higher-quality grades. However, certain grades of paper and board will still need to be made using a high proportion of new fibres as exemplified by surgical papers, and by grades in contact with certain foodstuffs.

The net effect of all these factors in terms of end use is shown in Fig. 7.1, which provides a 1993 analysis of RCF utilisation by paper grade for the European Union (EU) (excluding Nordic countries). While the overall average for RCF content is over 50% it can be seen that there are wide variations according to grade. Most RCF has been used in the lower paper grades such as packaging and board, where visual and printing properties are not so critical. Nevertheless, printings and writings show only 10% utilisation, and this segment is now receiving considerable attention, and it is here that much of the predicted growth in RCF utilisation is anticipated.

Recycling production processes and technologies

The overriding objectives of recycling waste are to maximise the recovery of reusable fibre and to minimise the foreign bodies known as contraries.

Contraries

There are three main groups of contraries:

- Paper mill additives, such as minerals, chemicals and functional aids.
- Conversion process additives, such as printing inks, plastics, foil, adhesives, and even staples and pins.
- Miscellaneous waste items, such as food, sand and plastic.

Typical problems caused by these contraries are summarised in Table 7.2. Thus it can be seen that the elimination of contraries is of vital importance in RCF and that it is a dominant factor. A typical process line contains many of the stages shown in Fig. 7.2 overleaf.

Table 7.2 Effect of contraries in waste paper

Impurity	Problem caused
Printing ink	Reduction in stock brightness, increase in speckiness, worsening 'stickies' problem
Adhesives	Blinding of paper machine clothing, reduction in equipment efficiency resulting in stoppage, reduction in product quality
Fillers	Reduction in quality of tissue products, reduction in yield and strength
Wet strength agents	Prevention of fibre separation at the pulping stage, loss of yield
Dyes	Reduction in stock brightness, problems with colour matching
Staples, paper clips, plastics, etc.	Damage to stock preparation equipment causing stoppage and reduction in stock and product quality

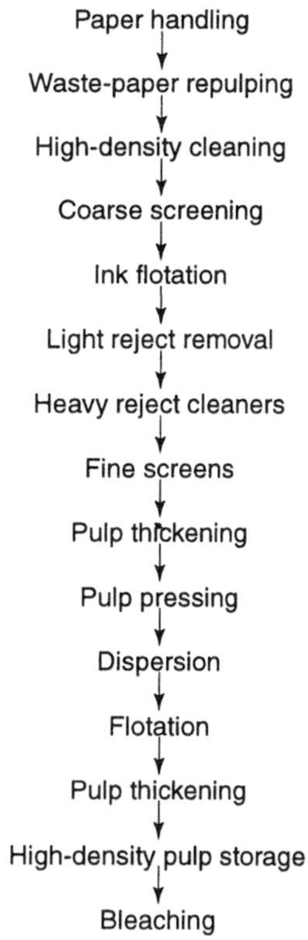

7.2 Recycled fibre process flow line.

De-inking

Emphasis has already been placed on the removal of contraries, but it is also worth highlighting the de-inking process, or the removal of ink and similar contaminants. Table 7.3 shows the market growth that has been predicted for de-inked fibre, where the use is expected to grow from 14 million tonnes in 1990 to 31 million tonnes by the year 2000. De-inked fibre consumption will therefore more than double in the decade and by the year 2000 it will account for at least 20% of total RCF.

A significant proportion of the extra de-inked tonnage will be used in newsprint (over 55% of the total). Printings and writings will overtake packaging and become the third largest application for de-inked fibre.

Table 7.3 Growth in world de-inked fibre utilisation

	1990		2000	
Grade	**Million tonnes**	**% total**	**Million**	**% total tonnes**
Newsprint	5.5	40	15.0	48
Tissue	3.5	25	6.0	19
Packaging	2.0	14	2.5	8
Printing and writing	2.0	14	4.5	15
Market pulp	1.0	7	3.0	10
Total	**14.0**	**100**	**31.0**	**100**

Flotation

De-inking is again a selection of washing and flotation techniques. In flotation, the first stage involves detaching the ink from the paper. The fibre is processed under alkaline conditions to bring about fibre swelling and to enhance ink release through saponification. Depending on fibre characteristics required for the end product, other chemicals such as sodium silicate, bleaching materials and chelating agents (which render metallic ions inactive, so that the ions cannot interfere with the desired process) are used. The second stage accomplishes the removal of ink from the fibre and it takes place in the flotation chamber. Incoming pulp is mixed with air bubbles which bring the ink particles to the surface. The resulting foam which contains the ink is then removed.

In designing de-inking equipment it is important to take into account the effect of probable changes in ink formulations. Ironically some of these changes, stimulated by environmental pressures, may give rise to problems in de-inking. A good example is the move in the printing of newsprint from solvent-based to water-based flexo inks. The problem with the water-based ink is that the resins used in the formulation are alkali-soluble. Thus on pulping under normal conditions the resins solubilise, leaving minute ink particles. These particles are below 5 μm in size, and as they are hydrophilic, flotation de-inking is not effective. Wash de-inking which is designed to remove small particles also does not work, as the water-based ink tends to stain the fibres, and the very small ink particles become entrapped in the fibre structure.

Chemicals in office copying, particularly from non-impact printing, make ink difficult to remove, and are causing the development of new de-inking techniques, including the use of enzymes to detach the print.

It is possible to increase the quality and therefore the usefulness of certain grades of waste paper by a process known as fractionation, which causes separation of long fibres from a waste mixture. The residual shorter fibres may be usable, for example, in the middle layer of a corrugated case. One of the weakest fibres is found in newsprint, and is only acceptable because a newspaper has a short lifespan. Thus recycled newspapers are used together with magazine waste to make more newsprint. By the same token it is not possible to make high-quality writing paper from recycled newsprint.

Bleaching

Brightness is one of the most important properties of paper, particularly in the production of high-quality papers. In de-inking therefore, close attention must be paid to bleaching. Brightness will be affected both by the presence of ink, and by impurities such as dyes in coloured paper and lignin in mechanically produced fibres which are used in some newsprint manufacture processes.

Bleaching of waste paper is not straightforward. All recycled fibres will have undergone at least one drying process which can adversely affect their ability to accept additional bleaching. Metal ions may be present and they can cause decomposition of bleaching chemicals. Besides this, the different fibre types with varying histories invariably found in waste paper will each have different bleaching responses.

Waste paper processing is expensive. Depending on the final quality required, which may attract additional chemical costs, and including the cost of waste collection and transportation from urban areas to pulp mills, some RCF pulp can be at least as expensive as virgin fibre.

However, capital expenditure levels are greatly in favour of investment in de-inking plants rather than pulp mills. An investment comparison would suggest that a bleached kraft pulp mill would cost US$600 million for 1000 tonnes per day compared with a de-inking plant costing US$40 million for a capacity of 500 tonnes per day.

It is interesting to note that in the USA it is common practice for the government to offer significant tax concessions to encourage the building of de-inking facilities, and this has led to the appearance of a number of merchant de-inking pulp mills.

Market features of waste paper, and the urban forest

There are a number of components in the waste paper and recycling business:

- Collection and recovery.
- Sorting and disposal of discards.
- Transportation.
- Utilisation.
- Residue/sludge disposal.

It is important to note that the term 'recovered waste paper' means what it says, waste paper recovered from the total paper consumed. The term 'utilised waste paper' means the percentage of waste paper that is actually used in the product. On a global basis of course the two terms must be the same (assuming no stockpiles). But in Finland, for example, 31.7% of the waste paper consumed domestically is recovered, but because the export market for paper is so large, this represents just 4.8% of the paper manufactured.

While there have been a number of waste-paper merchants handling some bulk grades – for example, mixed waste and kraft lined strawboard (KLS) – recovery has traditionally been in the control of the local authorities. Utilisation is in the hands of the paper industry. This is in turn significantly governed by the investment required and by consumer attitude and demand.

The term the 'urban forest' has been adopted to describe the vast quantities of waste paper that are generated each day in major cities. The analogy is a good one. Most pulp producers go to great lengths to ensure a sustainable supply of cost-effectively produced wood of the correct quality for use as a raw material. Forest management, with benefits both to the papermaker and to the environment, has become an integral part of the pulp-producing operation.

For exactly the same reasons, the urban forest will require similar investment and management. Because the life cycle of the urban forest is shorter, it will respond more quickly to management control. The municipal solid waste (MSW) generation statistics in the USA are interesting, and they emphasise the large amounts of paper and paperboard in MSW (Table 7.4). The progress in recovering the components of MSW is shown in Table 7.5.

The US paper industry is now moving up the waste recovery stakes, achieving 40% in 1995 and aiming at 50% by the year 2000. Once a de-inking facility is on line, a plentiful supply of recyclable paper is vital. Mills may therefore try to build up a stock, leading to conditions of temporary shortage.

Table 7.4 US MSW waste generation (1993)

Material	Mass (million tonnes)	% of total
Paper and paperboard	70	37.8
Yard trimmings	30	16.2
Plastics	17	9.2
Metals	15	8.1
Food waste	12	6.5
Glass	12	6.5
Wood	12	6.5
Other	17	9.2
Total	**185**	**100**

Source: US Environmental Protection Agency.

Table 7.5 US recovery rates for selected components of MSW

Material	% of total recovered	Amount recovered (million tonnes)
Aluminium packaging	53	7.5 (aluminium and steel packaging combined)
Steel packaging	46	
Paper and paperboard	34	23
Glass containers	25	3
Yard trimmings	20	6
Plastics packaging	4	0.7

Source: US Environmental Protection Agency.

The key force for RCF utilisation has been economics. It enables mills and countries without abundant forest resources to compete effectively with integrated producers of virgin fibre-based grades. The impact of RCF on the competitiveness of mills can clearly be seen in the Western European newsprint industry. Here the most cost-competitive production is based on RCF and is located close both to RCF sources and to final customers. This has led to industry restructuring and relocation. Similar developments are now taking place in North America.

There are other reasons. There is, for example, an urgent need to divert the growing mountains of waste from costly and limited landfill. This is a move further underpinned by legislation and by public concern over the environmental impact of landfill causing contamination of groundwater and gas emission.

Market structure and operation

Waste from commercial sources in Europe has been collected by waste-paper merchants, often held captive by big paper companies who have organised their business on waste-based products. It is interesting to see the collection of waste newsprint similarly organised on a local basis by a main producer of recycled newsprint. Strong competition for waste-paper supplies and high waste-paper prices are forcing mills to re-examine the way in which they buy waste paper, and to understand the entire collection system better. Nevertheless it is believed that the European collection of printing and writing paper from offices is underdeveloped in Europe, and lags behind the USA in this respect.

The global recovery of waste paper has been steadily increasing from a figure of 49 million tonnes in 1980 to over 100 million tonnes in 1994. This last figure represents 39% of paper consumption, which was 268 million tonnes in 1994.

The historical trends for recovery from the main regions of the world are shown in Fig. 7.3. It is interesting to note the improvement in the North American recovery figure since 1980, coming more in line with Western Europe.

North America, Western Europe and Asia dominate the picture as paper producers, and they are clearly major RCF users (Table 7.6), but Asia and Western Europe produce high waste utilisation percentages, no doubt because they have poor forestry resources, and so are big pulp importers. As far as RCF is concerned, the current North American consumption of 33% still lags well behind Western Europe (40%) and Asia (50%). This also explains the main trade flow of waste paper from North America to Asia.

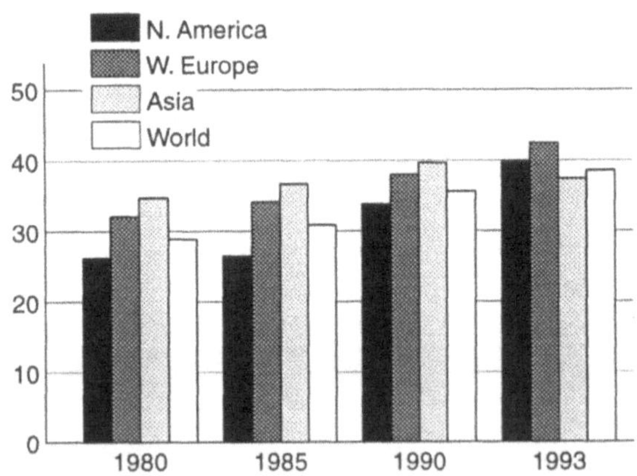

7.3 World-wide waste paper recovery (% of total paper consumed).

Table 7.6 World-wide RCF utilisation by region (1993)

Region	Paper production (million tonnes)	RCF utilisation (million tonnes)	%
Asia	65.8	32.9	50
Latin America	11.3	5.6	49
W. Europe	66.5	26.4	40
Africa	2.7	1.0	38
Australasia	3.0	1.0	35
N. America	94.1	30.8	33
E. Europe	8.3	2.2	26
World	**251.7**	**99.9**	**40**

On an individual country basis, the leaders in utilisation are:

- Taiwan 98%.
- Mexico 82%.
- Spain 70%.
- Netherlands 70%.
- Korea 69%.
- UK 60%.
- Germany 54%.
- Japan 53%.
- Switzerland 53%.

There are few other uses for recovered paper. The largest alternative use is in the manufacture of moulded fibre products such as egg boxes and fruit trays. Alternative uses are expected to expand, driven by the desire to divert material from landfill, and could include waste paper as a fuel.

Relative importance of RCF production

Both RCF and virgin fibre consumption will increase steadily over the next ten years, and this is clearly illustrated in Fig. 7.4. There is little doubt that recycling pressures, arising from further legislation and from environmental awareness, will mean that the use of RCF by the paper industry will continue to take a substantially larger and increasing share of the fibre supply. During the next ten years RCF utilisation is projected to grow at three times that of virgin fibre, i.e. 6% growth compared with 2%, and will result in RCF consumption

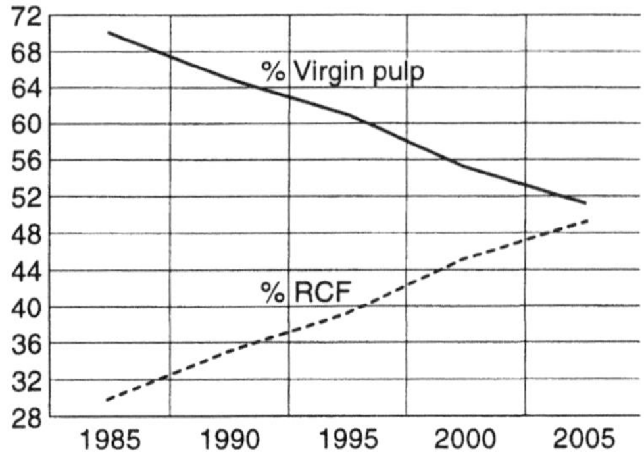

7.4 **World-wide growth of RCF utilisation by the paper industry.**

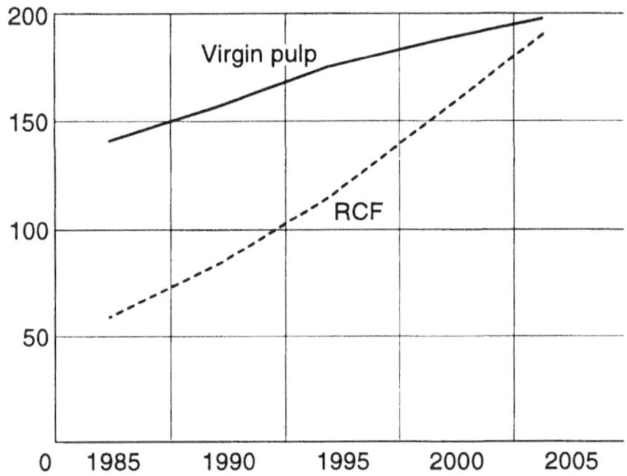

7.5 **Virgin fibre and RCF used (million tonnes) v. year.**

approaching the same volume as virgin pulp utilisation by the year 2005.

If the RCF consumption figure of 150 million tonnes is achieved by 2000, the utilisation rate will have increased from 35% in 1990 to 45% in 2000. Such a radical shift of this magnitude will cause the industry to make many changes, resulting in substantial restructuring of parts of the businesses. This is shown in Fig. 7.4 which indicates a 50/50 split in utilisation between virgin fibre and waste paper by the year 2005.

Figure 7.5 shows that over the decade 1995 to 2005, if the growth of virgin fibre utilisation is only 2% per annum and projecting these figures, an additional 76 million tonnes of RCF will be processed compared with an extra 22 million tonnes of virgin fibre. The rate of increase of RCF usage is greater than the rate of increase of virgin fibre.

It was estimated by Jaakko Poyry that the waste-paper disposal burden (the difference between paper consumption and waste-paper recovery) will grow by about 20%, or 32.9 million tonnes, over the period 1990 to 2005. This difference reflects the efforts to increase the recycling of waste paper, against a projection of paper consumption growing by 45% (3% per annum) during that time. In North America, the disposal burden will decline slightly, reflecting the dramatic moves to increase waste-paper recovery, while in Asia and Latin America the burden will grow significantly.

Nevertheless, despite all of these efforts, the disposal burden is worsening.

Legislative pressures to increase RCF utilisation

There is little doubt that environmental concerns backed by both legislation and certain national initiatives will continue to exert pressure to increase RCF utilisation further. They are principally seen in Europe and the USA, as follows.

The EU Directive on Packaging and Waste was adopted in 1994. It will have significant implications, originally giving member countries just 18 months to adapt their national legislation to meet these provisions, namely:

- Recovering 50% to 65% of packaging by weight within 5 years of implementation. For each material, 15% minimum must be recycled and there is a target range of 25% to 45% for recycling each material.

EU eco-labelling is another relevant piece of legislation:

- All products that are awarded the eco-label (so far, tissue and towelling) have been assessed according to seven parameters. These are:
  - renewable resources
  - nonrenewable resources
  - carbon dioxide
  - sulphur dioxide
  - COD
  - AOX
  - waste

  Points are allocated according to the level of each parameter in the product, and to obtain a label a product must not exceed a specified total load points score.

It is intended to review the criteria every three years, and if possible, raise them. The fundamental philosophy of the scheme is to encourage a progressive increase in the environmental standing of consumer goods.

EU eco taxes

The Belgian government tried to introduce a broad-ranging eco tax in 1993 on the sales of certain products, including most paper grades. The proposal was intended to apply to domestic production and imports, but not exports. The main objective behind the proposal relating to the paper industry was to increase the use of RCF, where the rate of recycling in Belgium was only 24%, the lowest in the EU countries. Unless criteria were reached, the proposal was to apply a tax of BFr 10 per kg of import.

By mid-1995 the tax had not been implemented and it is thought to be still on hold while further investigations are made with regard to the likely impact and possible contravention of trade regulations.

US legislative moves

It is interesting to review the US legislative moves. Consuming as they do 33% of the world supply of paper, they make the following approach:

- The 1993 Clinton Executive Order heralded a significant effort by government to promote recycling in the paper industry on a national scale. Fundamentally, the order seeks to increase market demand for papers containing RCF. A wide range of papers purchased by government departments was required to contain at least 20% post-consumer waste by December 1994, and the figure would increase to 30% by December 1998.

The effect would be significant as the Federal Government purchases about 300 000 tonnes a year of printing and writing paper, or about 2% of that market. RCF demand in the USA has been further stimulated by a number of government and industry initiatives, underpinned by actual or potential legislation (Table 7.7).

Over half of the states have introduced a mandatory 40% RCF content. If the same trend were to occur in Western Europe, this

Table 7.7 Pressure on US paper industry to increase RCF utilisation

Type of action	No. of states introduced
Specific recycling goals	40
Minimum recycling content	28
Tax incentives	18
Landfill bans	5

would present significant problems to Nordic producers who export a substantial part of their production to Europe. There the RCF utilisation levels are currently as low as 5% in Finland and 15% in Sweden. They could only be substantially increased by the importation of waste paper, and this seems to be at variance with the concept of sustainable, managed forests.

Environmental impact: energy

A study by the Austrian International Institute of Applied Systems Analysis (IIASA) raised serious questions about how far recycling should be taken. Broadly it found that maximum recycling increases the consumption of fossil fuels, which in turn increases emissions of sulphur dioxide and carbon dioxide. The de-inking process also consumes more energy than a self-sufficient modern kraft mill, and transportation of waste paper over increasingly greater distances adds to the energy consumption.

The IIASA does not imply that recycling is bad for the environment. It suggests there is an optimum level of recycling beyond which the advantages of recycling are outweighed by the disadvantages, giving rise to a negative environmental impact.

Waste paper is also regarded as a potential fuel. Scandinavian work has shown that bio-fuel waste-paper has a high calorific value and a relatively low environmental impact. The debate of recycling versus energy recovery is not new. Paper deemed unsuitable for recycling has been burned by the forestry industry boilers for some time. It appears to be a good substitute for coal and other fuels, particularly when it has been pelletised. It is anticipated that low-quality and non-recyclable waste paper will become a standard solid fuel for small-scale power production. The heat value is usually above that of bark which is 20 MJ/kg but below coal which has a value of around 30 MJ/kg.

Forestry investment

There is considerable concern about the way in which the use of increasing amounts of RCF could discourage investment in the forestry industry, and could militate against new plantings. It could also reduce the profitability of forestry operations to a point where they can no longer sustain forestry management activities.

Trade in waste paper

Global patterns

International trade in waste paper, of which there is a significant amount between North America and the Pacific Rim countries, is worth US$2 billion a year. The traded volume of global waste-paper exports is 15 million tonnes a year, which is about half the volume of pulp exported globally (29 million tonnes in 1994), and a quarter the volume of paper exported (71 million tonnes in 1994).

The importation of waste on a regional basis is quantified in Table 7.8.

Table 7.8 World-wide waste-paper import by region

Region	Waste-paper imported Million tonnes	% of total
Asia	6.0	39
W. Europe	5.9	38
N. America	1.6	11
Latin America	1.3	9
Other	0.4	3
World	15.2	100

Five countries import in excess of 1 million tonnes a year. It is interesting to note that this list is headed by two Pacific Rim countries.

- Taiwan 10%.
- Korea 10%.
- Canada 10%.
- Netherlands 8%.
- Mexico 7%.

Table 7.9 World-wide waste export by region (1993)

Region	Waste-paper exported	
	Million tonnes	% of total
N. America	5.6	37
W. Europe	5.0	33
Asia	0.5	3
Other	4.2	27
World	15.3	100

Waste paper exported by region is concentrated mainly on North America and Western Europe, as shown in Table 7.9.

There are relatively few countries with the ability to export significant quantities of waste paper. The lead countries are shown below, with their proportion of the world total:

- USA 35%.
- Germany 15%.
- Netherlands 6%.

The two main exporters of waste paper, the USA and Germany, account for about 50% of the world exported waste paper. When the US consumption reaches 50%, and this is targeted to occur in the year 2000, their ability to export is expected to diminish significantly.

Waste-paper pricing

The price of waste paper depends on a number of factors, but as with other commodities, the balance of supply and demand will have the biggest influence. Prices vary widely according to grade, as illustrated by the prices in Table 7.10 prevailing in Chicago in the middle of 1995 (seen in a copy of *Pulp and Paper Week*). The table clearly shows the range of prices per tonne from US$650 (hard white envelope cuttings) to US$135 (mixed paper). Prices exhibit similar trends to the sharp cyclical pattern of pulp prices, possibly being more volatile. In 1995 pulp prices rose to an all-time high, eventually topping US$1000 per tonne, before falling back.

Since 1993, export waste-paper prices have risen dramatically. In addition to the demand arising from the buoyant state of the paper industry, a number of other factors put pressure on prices. In 1994

Table 7.10 US waste-paper prices by grade fob seller's dock, June 1995

Grade	US$/tonne
Hard white envelope cuttings	650
Hard white shavings	560
Computer printout	460
White ledger	430
Sorted coloured ledger	320
Coated book stock	310
New double lined Kraft	260
White blank news	260
Corrugated containers	180
News	170
Mixed paper	135

Germany moved to reduce waste paper collected by 15% in order to stabilise prices and reduce loss- making exports. Other big influences have been the start of major newsprint capacity based heavily on waste paper, together with a high demand from the Far East, and a general move to build up stocks following low levels in 1992/93 when supplies were plentiful.

The upward pressure on waste-paper prices may well remain in the long term. Main exporting countries will continue to need more waste paper domestically in order to meet ambitious utilisation targets. This could lead to shortage for export followed by a price escalation. With timber shortages also predicted because of environmental concerns imposing constraints on logging, virgin pulp prices could also be under pressure.

Paper recycling reaches record levels in Europe

It is of considerable interest, after the discussion in this chapter, to quote from the August 1997 press release on the subject, from the Paper Federation of Great Britain. It contains the following information.

A record 32 million tonnes of paper were collected and reprocessed into new paper in Europe during 1996. Figure 7.6 shows the progress in increasing the amount of recovered paper utilisation in Western Europe since 1991. The recovery rate is the amount of recovered paper collection and subsequent utilisation expressed as a

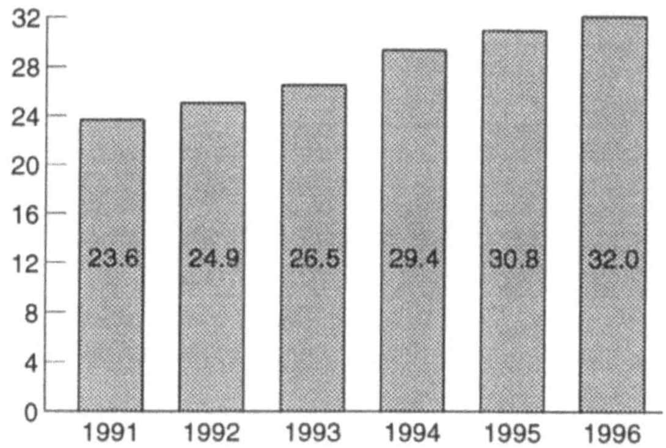

7.6 Recovered paper utilisation in Western Europe (million tonnes).

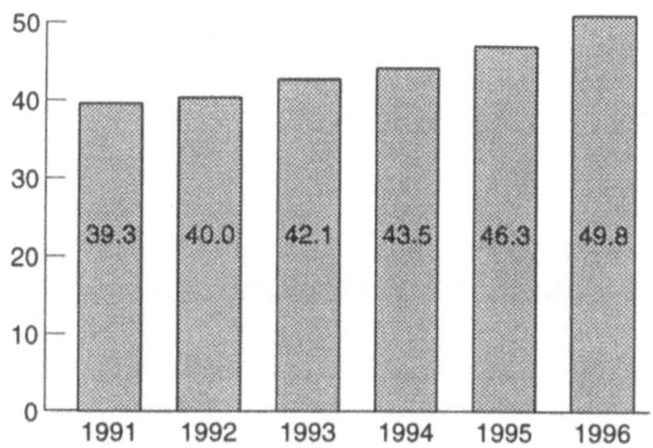

7.7 Per cent utilisation in Western Europe.

percentage of total paper and board consumption. In 1996, the European paper industry achieved a recovery rate of 49.8%, an increase of 3.5% over the previous year. According to the Confederation of European Paper Industries (CEPI), paper recovery has grown from 39% to the current 49.8% in only 5 years. The figures are shown in Fig. 7.7. Comparable figures for the USA, Canada and Japan are 41%, 45% and 51.3% whilst the world average is 37%.

Paper recycling has increased in all European countries. The present recovery rate is one of the highest of any industry in the world. 'Paper recycling is a technical, economic and environmental reality' said Dr Holzhey, chairman of CEPI's Recycling Committee. 'Citizens have played a key role in collecting and segregating used paper. They know how to distinguish what is waste from what is a valuable resource. The industry produces new paper products using this raw material which would otherwise end up in landfill. For the paper industry this is not the result of regulatory measures or impositions. For us the attractions are obvious and the growth has developed naturally. Used papers are not waste, and both the public and industry know it.'

'Europe is self-sufficient in recovered paper and there is an overall balance between the amount of paper collected and the industries' capacities for recycling. However the situation is not the same in all European countries. The use of recovered paper is most attractive for papermills situated in densely populated areas where large supplies can be transported over short distances, making the process economically viable. Countries with rich forest resources provide the high quality virgin fibres that are required to maintain these high recycling rates.'

'To optimise paper recovery in Europe we require free movement of collected material. Almost 20% of the paper recycled in Europe has to cross a border.'

'There are limits to recycling. If it becomes too expensive, the whole system would collapse. Also some papers like hygienic papers and wallpaper can never be recovered. There are also technical limitations. Each time the paper is recycled the fibres become worn, and some are destroyed or lost. Paper production cannot solely be based on recycled fibres. A functional, environmental and economically sustainable paper cycle requires the continuous injection of primary fibre as well as an intense recycling loop.'

'Everything points to the conclusion that paper recovery in Europe can be increased further given the right conditions. 'A high level of recovery demands that all players fulfil their share of responsibility. Individuals and companies have to participate voluntarily in collection systems. Local authorities have to provide adequate facilities and infrastructure. The paper industry has to invest to create greater capacity. And finally, legislators have to establish the necessary framework in which this loop can run naturally, without impediments or artificial distortions.'

Summary

The classification of waste paper and recycled fibre has been described, and an outline of the recovery processes shows the steps to be taken to make RCF suitable for papermaking. Contraries in the waste are identified, showing the problems they can cause the papermaker.

A recent report (1996) prepared for the European Commission by the Institute for Prospective Technology Studies (IPTS), exploring the impediments to recycling in Europe summarises the following key examples:

- Presence of complex products, such as multilayer packages.
- No stringent packaging legislation in some countries.
- Generally insufficient waste supply in Europe, even if the picture can vary according to the grades of paper considered. Supply is usually short in the higher-grade papers.
- Instability of waste-paper prices: it appears that small imbalances in the supply/demand situation result in strong price swings. One of the main reasons is seen in the lack of information, and the improvement of statistics is proposed as a solution.
- Economic competition from incineration.
- Large amount of paper sent to municipal solid waste (paper mixed with municipal solid waste cannot be recycled).
- Cost differentials and subsidies (or lack of) distorting the competition between disposal options. Whenever landfill is a cheaper option there is little incentive to collect and separate paper. The different ways of subsidising quality paper collection in different countries also causes problems.
- Lack of efficient collecting systems and expensive sorting.
- Recycled material content targets can be counter-productive if set arbitrarily. Recycled paper should be used for its best applications.
- Unequal distribution of reprocessing capacity.

World waste figures have been considered together with the impact and implication of the urban forest. Legislative measures to increase RCF utilisation were also considered. The progress made to legislative waste usage was considered as was the relationship to the sustainable forest concept. World utilisation statistics showed significant trade in waste paper between the North American and Pacific Rim importers.

CHAPTER

8

Environmental issues

Public perception and factual analysis are not the same thing, and the chapter commences by reviewing one or two environmental 'statements' in the light of the pulp and paper industry in an endeavour to put the environmental aspects of the industry into perspective. It considers the steps being taken to improve the environmental aspects of forestry and to improve pollution control. The recycling directives, and particularly their application in Germany, are considered, as are the voluntary schemes such as BS 7750 and the ISO 9000 series, the European Union Management Standard (EMAS) and the Comité Européen Normalisation (CEN), the European standardisation body.

Public perception

In earlier chapters we have dealt with the manufacturing processes and with forestry, raw materials and recycling. Each time, some of the environmental concerns were discussed. This chapter is concerned with bringing together the principal environmental issues. It may be perceived that a great number of the arguments about the environment are real, although some of them are virtual. The following quotations from current literature support this view:

> *From a scientist's viewpoint the rapid changes of values is challenging and interesting. From a politician's, they are confused. They require straightforward answers – but there are no simple rules for the environment any more.*

Under the headline 'World getting warmer – but at slower rate' the *Daily Telegraph* on 22 January 1997 had the following report:

> *Global warming is continuing despite cooler temperatures in 1996, according to figures released yesterday. Last year's average temperature was 0.21 °C warmer than the average over the past 30 years of 11 °C. The change in 1995 was 0.38 °C. However, 1996 was still the eighth warmest year since records began in 1860 according to the Met Office and the University of East Anglia.*

The Friends of the Earth in 1987 estimated that doubling the carbon dioxide in the atmosphere will cause an average global temperature rise of 3 °C. In 1850 there were 265 ppm of carbon dioxide in the atmosphere; in 1985 there were 340 ppm: the projection for the year 2050 is 600 ppm.

Perhaps in all of these situations one should pause and remember Newton's Third Law of Dynamics which says that 'For every action there is always an equal and opposite reaction'. The factual statements are correct, but the conclusions are – as they can only be – based on the available facts, and in talking about long-range effects it is necessary to take account of other information. For example, cutting down forests has an environmental down-side, but cutting down forests in a managed way ensures that there is a good supply of young trees to promote the better exchange of carbon dioxide from the air and its replacement with oxygen. So that on the one hand paper made from pulp may be considered to be environmentally bad, but on the other hand it is ecologically sound.

Consider a news flash released to the British media in January 1997, regarding the flotation of a new company. This caused brochures and papers to be sent out, using numerous tonnes of paper. The Company says it is planting trees to restore the environmental balance. Had there been a better understanding of the world scene, there would not have been misleading inferences taken from it. For example, Friends of the Earth suggested using less paper so that the problem did not arise. But some rational, informed thought says there is no problem. The forests need to be a managed, sustainable resource – and they have to be if the industry has regard to its future wood supply, as it does. Consequently the industry, more often than not, has a management programme that involves regeneration.

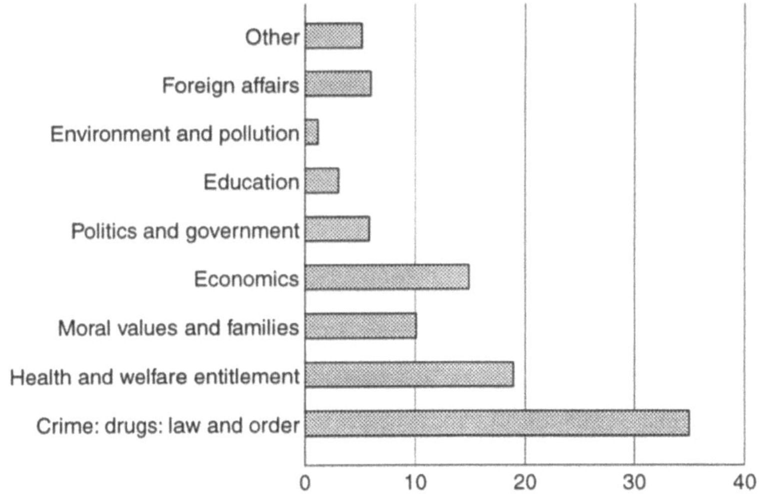

8.1 The single most important issue in the USA. (Source: US population survey 'Policies and public policy', Market Strategies Inc, 1994.)

When a group of Americans were surveyed in 1994 and were asked 'What do you think is the SINGLE most important problem or issue facing the country (USA) at the present time?', the poll produced the results in Fig. 8.1, which give an indication that environmental issues are no longer at the top of the public's concern.

It is unlikely that this is a real attitude shift in the public's mind, coming about through them being better informed. Instead, it is more likely that the public perceives that the issues are now being taken care of through integration into government policies and the forward plans of corporations, where the issues are of the greatest importance in the corporate mind. One way or another, the problem has been taken on board.

There are a number of ways forward, and they embrace emotion, cash return, regional and local issues, company self-interest marketing and green-field site developments. But the most important fact is that on such an important issue as this, we must get to the stage where the decisions that we take are logic-driven and not emotion-driven.

Perhaps the last word on public perception should be given to David Clark, speaking in the Eucepa Symposium Recycling '95, hosted by the UK Paper Industry Technical Association:

> ***the industry will have to convince the public of its clear commitment to environmental improvement, demonstrate its progress on sound science grounds and above all, communicate both its commitment and progress. The industry can no longer take for granted access to its global***

market or to its source of raw material, primary or secondary.

Over the past few years, a number of major political changes have occurred around the world which inevitably have a radical effect upon environmental policy. The swing of power in the United States has put in doubt some of the ambitious green plans of the first Clinton Administration. The entry of the Nordic bloc (Norway, Sweden and Finland) into the European Union with close to 60% of West European pulp production, will have particular implications for the paper industry, among them the fact that the degree of self-sufficiency in wood raw material in the EU will rise significantly.

Some European countries, particularly the UK and Germany, have set up working groups to campaign against over-regulation by Brussels. They point out that this can be a dangerous and a growing impediment to growth and competitiveness.

Sustainable forestry

The conservation of forests remains a priority item on the agenda of the Green movement, as shown by recent campaigns against certain forest practices. There seems to be agreement from most sides that certification of the forests in one form or another is inevitable. The WWF has stated that the FSC (Forest Stewardship Council) forest certification programme is now underway with 4 million ha of forest certified world-wide. Although this represents a very small proportion of the total forestry base, it enables the WWF to claim that the FSC concept can be made to work. Not surprisingly parts of the timber industry are not comfortable with attempts to set these fundamental standards.

The further debate will be directed at:

- How schemes developed at regional, national and international levels can be harmonised.
- Determining who will bear the cost of sustainable management and the certification process itself.
- Whether plantations will be rated more favourably than natural forests which could have a radical effect on future virgin fibre sourcing policy.
- Finding the means of tracking wood products through the many stages from forest to consumer.

Table 8.1 Resolutions of General Declaration of the 1993 Helsinki Conference

H1.	General guidelines for the sustainable management of forests in Europe.
H2.	General guidelines for the conservation of the biodiversity of the European forests.
H3.	Forestry cooperation with countries with economies in transition.
H4.	Strategies for a process of long-term adaptation of forests in Europe to climate change.

Companies are still uncertain about the precise benefits, but it is being suggested that certification could lead to advantages such as increased market share, better company profile and less risk from consumer boycotts.

World-wide there are a number of international government initiatives in progress to reach consensus on indicators of sustainable forestry. These include schemes from groups as diverse as the Montreal process (non-European), the Helsinki process and the International Tropical Timber Association.

As far as Europe is concerned, the 1993 Helsinki Ministerial Conference on 'The Protection of European Forests' produced some broad guidelines for sustainable forestry which are still relevant. All the governments of the Confederation of European Paper Industries (CEPI) member countries, the European Commission and other European countries are signatories of the General Declaration of the Conference that tabled four main resolutions. These resolutions are shown in Table 8.1. The follow-up Geneva meeting in June 1994 resulted in the adoption of sustainable forest criteria which were deemed to be helpful in developing the process of certification.

CEPI believes that future operational forestry principles and emerging certification schemes should be heavily geared to locally acceptable practices with linking and harmonisation on an international level.

One factor that has significant effect on policy making in the forestry sector is the ratio of public to private ownership. This varies widely across the pulp-producing countries (Fig. 8.2). In Canada, with 94% public ownership, the industry is pressing for a national standard on forestry certification with the hope that this will encourage the ISO to adopt it as a starting point for drawing up an international standard.

Finland, with 29% public ownership, is at the other end of the scale. With over 400 000 private forest owners, implementation of certification would inevitably present some administrative difficulties. It would also be costly to some small operators, as opposed to the big companies with a major share.

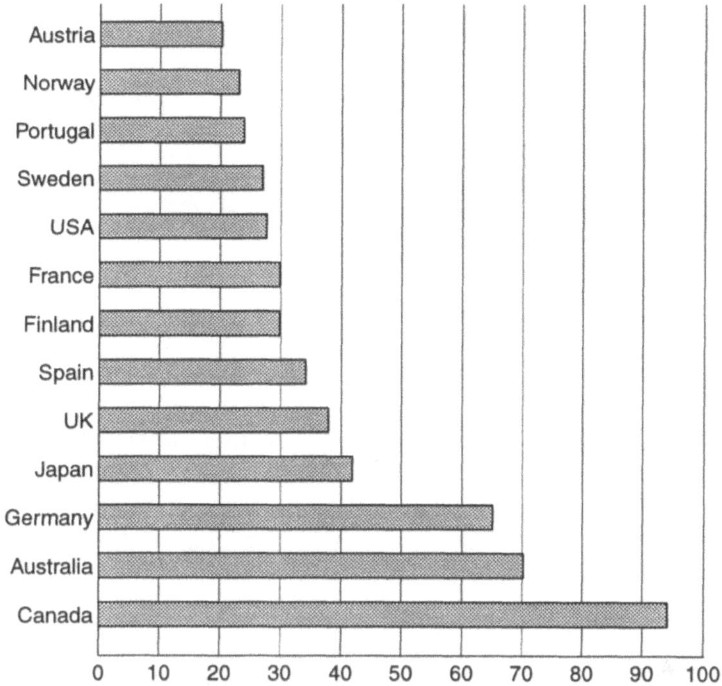

8.2 Public ownership of forest land in selected countries (%).

In the USA where private ownership is more than 70% there are signs of a significant backlash against curtailment of wood harvest, of the sort that has occurred in the Pacific Northwest, because of environmental concerns. Many US landowners are insisting that they be allowed freedom to exercise their basic land rights. A nationwide survey (Project Common Sense) was carried out in March 1995 (Fig. 8.3) to determine how the public felt about the enactment of the current form of Endangered Species Act (ESA). The survey concluded that an overwhelming majority of the US public rejects the way the current ESA restricts private forest owners. Instead they believe that establishing incentives and reward for landowners and land users would be a more effective approach to protect species.

There are real signs that some of this protest is getting through to the government administration. In March 1995, the US Senate passed legislation that placed a six month moratorium on the listing of endangered species. This was designed to hold back any further restrictive action until Congress has had the opportunity to examine the full impact of the ESA.

While participation in forestry certification is generally considered to be voluntary, some advocates of certification, particularly in

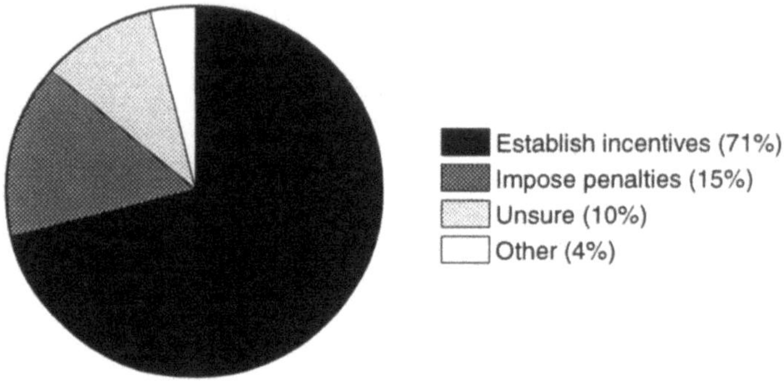

8.3 Public survey: most effective methods of protecting endangered species.

the USA, have been attempting to get certification passed into law. Already the states of New York and Arizona require certification for the importation of tropical timber. There are signs that the same could happen in Europe, where a number of countries are trying to introduce laws requiring the labelling of tropical timbers.

Pollution control

It is interesting to trace the development of the underlying concept for integrated pollution control in the UK. Here the launch of the Environmental Protection Act in 1990 signalled the enactment of a new phase in environmental control. A major step in 1991 was the introduction of integrated pollution control (IPC), whereby a wide range of industries including the paper industry were prescribed as potentially polluting and identified for control. IPC, with its emphasis on the use of Best Available Techniques (BATs), represents a more holistic approach directed at minimising the pollutants produced by the prescribed industries. It could well be an extremely powerful tool for bringing about environmental improvement.

In September 1993, with a watchful eye on the UK experience with IPC, the European Commission adopted a proposal for a Council Directive to be known as integrated pollution prevention and control (IPPC). This concept, similar in many respects to the UK IPC system, will introduce a system of integrated permits which will be required as authorisation to continue operating. It will cover emissions to all environmental media - air, water and land.

IPPC is now seen as an important element of the overriding EU objective to achieve sustainable development. The original intention was that member states translate the directive into national legislation by June 1995 with existing plants caught by the directive being required to meet the appropriate standards within ten years. This is now in place, just a little later than was originally planned.

The benefits that could result from compliance with IPPC have been put forward as:

- Savings from minimisation of releases.
- Greater environmental awareness for management and operatives.
- Limitation of potential liabilities.
- Exposure to newer and more efficient technology.

As far as the paper industry is concerned, without doubt, mills will be forced into a deeper understanding of the environmental consequences of all their operations, particularly with regard to emissions, use of resources and process management. Inevitably the pressure created will be transferred back to the suppliers of raw materials and equipment.

One of the key controversial elements of the scheme is related to emission limit values (ELV). There has been fierce debate about the parameters that will be used to set these emission values. One school of thought prefers BATs that could be applied uniformly across Europe. The other group favours Environmental Quality Standards (EQSs) which set standards for whole areas, such as the ability of fish to survive in a given body of water.

There seems little doubt that in the future, pollution control of water will be more and more stringent and will include AOX and also air pollution control. It will also lead to landfill becoming either more expensive or completely banned. New measures will be introduced relating to the toxicity of heavy metals and to dioxins.

As legislation tightens, so water reuse becomes more important. It therefore makes sense to recycle water, and by so doing, reduce consumption by using whitewater (the industry term for process water that is recycled within the process) and not freshwater, and generating freshwater from whitewater. The zero effluent system will become a reality.

In the USA, there were 8000 landfill sites in 1988 which by 1994 had been reduced to 3558. Unfortunately this does not point to the success of the initiative, but rather to the expense of landfill and the difficulty of getting more landfill approval because of cost and adverse public opinion.

The packaging directive

The EU Directive on Packaging and Waste was finally adopted in December 1994. This will have far-reaching implications for many sectors, including manufacturers, distributors, local authorities and retailers. The Directive gave member countries 18 months to adapt their national legislation to meet the provisions.

Through this directive, countries were committed to recovering between 50% and 65% of packaging by weight within five years of implementation. For recycling the target range was 25- 45% by weight with a minimum of 15% for each material. The upper limit for recycling was set to correct the severe distortion of the waste materials markets that have developed in the EU over the past few years. In particular, Germany has been flooding neighbouring countries with its surplus of waste materials collected under its strict 1991 Packaging Ordinance.

Not surprisingly, this EU Directive has had a turbulent passage since it was launched in draft form by the European Commission in 1992. The three countries, Germany, the Netherlands and Denmark, who have already achieved high recycling levels, called for the removal of the upper limit and threatened to challenge the legality of the Directive.

The German national targets originally programmed to be met by July 1995 may cause some recycling levels to be restrained by the upper limits for recycling contained within the new directive (Fig. 8.4). Meanwhile, some member states insist that an upper limit must be maintained if viable markets for recycled materials are to be developed. The directive's final wording gives the European Commission powers to grant allowances to countries wishing to exceed the upper limit, but only when there will be no adverse effect on other states.

A German view on this was expressed by Professor Lothar Gottsching in his significant paper 'The German Scene and the influence of legislation' given in the Eucepa Recycling '95 symposium:

> *Germany believes in the necessity and feasibility of perfectly controlled waste management in a recycling minded society and in an industry aiming at sustainable development.*

Gottsching lists the driving forces for waste utilisation as:

- Economy of recycled fibres (first and globally valid priority).
- Ecological factors (medium priority, e.g. energy saving v. mechanical pulping).

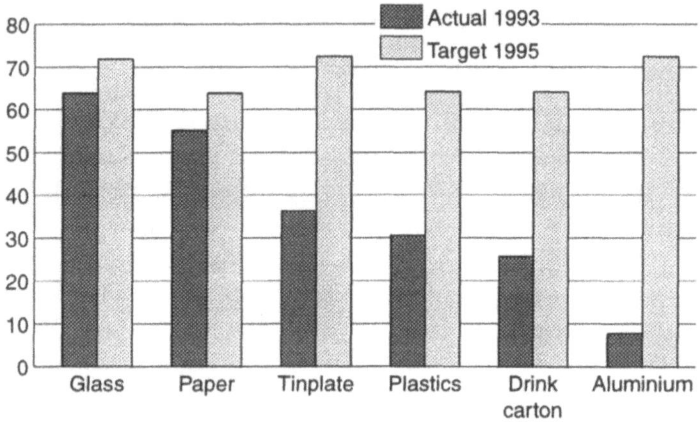

8.4 Restraints on German recycling by EU packaging waste directive.

- Legal requirements (globally low priority but locally high - Germany and the USA).
- Eco marketing.

In Germany the Waste Paper Management Act 1986 focuses on:

- Avoidance of waste.
- Recycling waste.
- Environmentally friendly disposal of waste.

The packaging ordinance 1991 requires rapid recovery and recycling of 100% pre-consumer packaging waste (packing needed to protect the contents during transportation and to ensure its undamaged arrival) and 60% post-consumer waste (sales packaging used to present a good image of the product to the customer).

The 1994 German Waste Avoidance, Recycling and Disposal Act deals with recycling and waste management - the product is the responsibility of the manufacturers and the distributors, and they must be aware of their responsibilities in the disposal of any waste that arises.

Gottsching goes on to discuss voluntary agreement for recycling targets and a recycling level of 60% for printed matter and office waste. He says, 'Considering the goals the approach of the Voluntary Agreement can be regarded as liberal because it does not force the paper industry to produce paper and board with set waste paper utilisation rates'.

The Dual System Deutsche (DSD) demands sales packaging recovery by paying a licence fee for an agency for package producers (mills, etc.):

> *The paper recovery business is now not supply demand, but supply only through legislation. This is the key to reduce the volume of solids to be disposed of by landfill.*

Finally, Gottsching lists the German Technical Guideline – at least by 2005, no solids with greater than 5% organic matter can be land-filled, therefore they must incinerate them if possible within the industry energy generation requirements. This has the benefit of utilising excess waste paper through energy recovery which shows a saving on fossil energy and which produces less greenhouse gasses, so reducing the greenhouse effect.

Voluntary schemes

In addition to legislation, a range of voluntary schemes is being developed. These can be split broadly into two groups:

- Standards governing how a company manages its business which calls for a certain level of competence in environmental management.
- Regulations dealing with the impact of the product on the environmental scene, e.g. eco-labelling.

Environmental Management Scheme (EMS)

Running a successful business requires sound management with clearly defined goals and strategies. The application of ISO 9000 and BS 5750 to the management of total quality is now well established in many companies.

Systems are now being proposed whereby management of environmental performance becomes an integral part of the overall company plan. Whereas IPPC is primarily concerned with pollution control, the EMS concept requires the organisation to identify and evaluate all the environmental effects of its activities.

Possible benefits that may be derived from the introduction of an appropriate environmental management system are identified as:

- Credibility – a positive sign to all stakeholders that the company is committed to minimising environmental impact from its activities. Many banks now seek reassurance of a company's environmental record before making loans.

Table 8.2 Environmental management systems

Standard	Level	Status
BS 7750	UK	Established in early 1995
ISO 14001	International	Adopted
EMAS regulation	European	Established in April 1995. Deals with producer responsibility for waste, forest certification and eco-labelling.
CEN standard	European	Will come after the development of ISO 14001

- Competitive advantage - environmental requirements are increasingly being built into procurement activities and the presence of some environmental management may give an advantage when competing for contracts.
- Risk - exhibiting good risk control, of which environmental management is a part, may result in reduced costs, for example from lower insurance premiums.
- Liabilities - environmental pollution can now result in penalties both for companies and for management personnel. A management system which identifies environmental impacts should help minimise concern and liability.
- Compliance - close monitoring of the operation should reduce the risks from non-compliance.

Currently a range of environmental management schemes (Table 8.2) is being developed around the world. Some confusion is arising as to which of these schemes any individual company, particularly those operating internationally, should aim for. In the UK, the Confederation of British Industry (CBI) has produced guidelines to help companies decide which standards are most applicable for them and to provide signposts for the future.

Fears that the emergence of different environmental management standards for plants in Europe and the USA could create a barrier to trade have subsided following a meeting of the standardisation bodies in November 1994. BS 7750 is now accepted by the European Commission as specifying a management system for EMAS and for the ISO 14000 series.

BS 7750

The UK has again emerged as a leader in developing an environmental standard. The concept of a British Standard for environmental management was first launched with much enthusiasm in 1991. Then the draft BS 7750 was produced in record time with a full version available early in 1992. However, some momentum was lost and it was not until the spring of 1995 that companies were able to seek accredited certification to BS 7750. It had been widely envisaged that BS 7750 would form the basis of any European (CEN) standard and that it would satisfy the management system requirements of EMAS regulation. However, it now seems more likely that CEN is moving towards adopting ISO 14001. If this happens it would mean the end of BS 7750 in its current form, and its replacement by an ISO/CEN standard. In practice, as BS 7750 is more demanding than the ISO standard, any company who had implemented a system based on BS 7750 would be unlikely to have any difficulty in complying with the ISO/CEN requirements.

Meanwhile, in the absence of a CEN standard, equivalents of BS 7750 have been introduced in Ireland, France and Spain.

Whatever form the final standard takes, it is likely that most mills would be wise to install an environmental management system of one form or another. Experience in other industries has shown that they are a key component in operating successfully within the IPPC type regime.

ISO 14001

The ISO is developing a package of environmental standards known as the ISO 14000 series. While most provide general guidance, ISO 14001 is a certifiable standard containing specifications for an environmental management system.

The basic specification is centred on the same core management system elements as BS 7750, that is, plan, implement, check and review. It does not impose any additional requirements beyond those of BS 7750. The steps in applying BS 7750 are shown in Fig. 8.5.

ISO 14001 reflects a compromise between the European concept based largely on BS 7750 and EMAS, the less prescriptive North American approach.

Unfortunately, recent developments in ISO 14001 may cast some doubts on the uptake of EMAS. It is suggested that strong lobbying by certain US business sectors have made the latest drafts of the ISO standard considerably weaker than both BS 7750 and EMAS.

Commitment
↓
Initial review
↓
Policy statement
↓
Organisation and personnel
↓
Register of regulations and their effects
↓
Set objectives and targets
↓
Management programme
↓
Management manual
↓
Operational control
↓
Records
↓
Audits
↓
Management review

8.5 Management system for applying BS 7750. (Source: PT 33:11 pp 16 WJ Inman)

EMAS

In March 1993 the Council of Ministers adopted a regulation that allowed companies to participate in a community wide eco-management and auditing scheme to be known as EMAS. The system as a whole was launched in April 1995.

EMAS enables a company to demonstrate that it has been able to meet a number of environmental requirements to the satisfaction of an independent environmental verifier accredited by the member state in which the scheme is in operation. In order to be accredited, the participating company must demonstrate that it is committed to establishing and implementing management systems designed to protect the environment.

When the scheme was introduced in the UK, the Environment Minister said that it would enable companies to cut costs and gain international competitive advantage. He suggested that registration under EMAS will also provide further advantages, namely:

- Superior corporate reputation.
- Enhanced customer relations.
- Improved profitability.

The key difference between EMAS and standards such as BS 7750 and ISO 14001 is that in order to obtain EMAS accreditation, the company must also be prepared to provide information on the environmental effects of the operation by issuing an environmental statement. Participating companies would be subject to a series of audits, to be made public, which must be undertaken at intervals of no less than three years.

Those within the industry will have to balance the competitive advantage in the market-place to be gained from EMAS against the requirement under the scheme to allow themselves and their operating procedures to be open to public examination.

In the meantime, the European standardisation body CEN will consider whether to adopt ISO 14001 as the European standard for environmental management within EMAS. The overall requirements of EMAS are stricter than those of ISO 14001. Therefore, companies exporting to the EU fear the adoption by the EU of its own management standard under the EMAS regulation might give certified EU companies an unfair advantage.

Cefic (the European Chemical Association) wants its members to register under EMAS because it is the Director-General XI's first attempt to promote a voluntary agreement with industry, instead of imposing traditional command and control legislation.

The regulation is subject to review in 1998 and there is a body of opinion that would like to see the scheme becoming mandatory in certain circumstances. In the meantime it is interesting to see a listing of European Paper Companies using an EMS system (Table 8.3).

Eco-labelling

The subject of eco-labelling has now become controversial within the paper industry. The matter was recently brought to a head when two of the European paper industry associations, CEPI and European Tissue Symposium (ETS), publicly announced severe doubts about certain aspects of the scheme for paper products.

In the EU, the detailed criteria for tissue (toilet paper and kitchen rolls) were agreed at the Regulatory Committee early in 1994 and were formally adopted by the commission in November 1994. The expectation was to have labelled products in this category in the shops by mid-1995. This was slow to materialise, finally starting to appear in 1996.

Table 8.3 European companies with EMS schemes

Country	Company	Location	Scheme
Austria	KNP Leykam	Gratkorn	ISO 14001
	Neusiedler	Theresienthal	ISO 14001
	SCA	Laarkirchen	ISO 14001
Belgium	Scott	Duffel	BS 7750
Finland	Enocell	Ulmaharju	BS 7750
	Metsa	Kemi	BS 7750
	Metsa	Aanekoski	BS 7750
	UPM/Kymmene	Kuusankoski	BS 7750
	UPM/Kymmene	Pietarsaari	BS 7750
Germany	Feldmuhle	Hagen	EMAS
	Strepp	Kreuzau	EMAS
Italy	Favini	Vicenza	EMAS
Norway	Norske Skog	Skogn	ISO 14001
	Peterson	Moss	EMAS
UK	Arjo Wiggins	Dover	BS 7750
	Chapman	New Waterside	BS 7750
	Curtis	Dalmore	BS 7750
	Curtis	Guardbridge	BS 7750
	Inveresk	Carrongrove	BS 7750
	UK Paper	Donside	BS 7750
	UK Paper	Grovehurst	BS 7750
	UK Paper	New Thames	BS 7750
	UK Paper	Sittingbourne	BS 7750

Source: The Green Paper, David Price Forest Products, 1996.

The next stage in the EU eco-label paper product programme was to develop the criteria for fine papers, and this has been carried out. For the paper industry, this is a crucial step for a number of reasons:

- The printings and writings sector accounts for close to 40% of all European paper production.
- This sector uses by far the lowest proportion of recycled fibres - less than 10% compared with the overall industry average of about 40%.
- The largest application for market pulp, where brightness, chlorine bleaching and AOX are key issues, is in this sector.

The final *ad hoc* European Union eco-label fine papers meeting, run by the Danish Environmental Protection Agency as the lead working group, took place in Copenhagen on 6 April 1995. Representatives were present for the European Commission, industry environmental groups and national eco-labelling authorities.

The proposals for fine paper criteria put forward were similar to those for tissue papers with the following exceptions:

- Renewable resources (RR) – the scoring for RR has been dropped. This is because fibre can only be recycled six or seven times and fine paper is the best grade of paper in which to introduce virgin fibre.
- The weighting factor for waste has been reduced because the scoring for tissue papers gives a strong encouragement to recycled fibre. As the meeting could not agree on this point, it will have to be resolved at a separate Competent Bodies meeting.
- The hurdle for AOX has been doubled – that is 1 kg/tonne paper compared with 0.5 kg/tonne paper for tissue.

In view of the pressing need to finalise this label, a number of other outstanding points were put on hold and tabled for investigation during the three year revision period. Those included items such as the environmental impact of fillers (up to 20% of paper weight), full guidelines for sustainable forest management, alternative to AOX test and the use of chemicals.

CEPI

In April 1996, an announcement was made by CEPI that it would be withdrawing from any further involvement in the development of criteria for the EU labelling of fine paper products. Whilst recognising that the papermaking process does have an impact, CEPI states that it does not believe it possible to eco-label a process, in the way the present scheme attempts to carry it out.

Instead CEPI sees EMAS as the best way forward as this was designed specifically to achieve continuous environmental improvement in the process which it says is what is fundamentally required. Here the entire industry could aspire to it rather than the top 10–20% of paper products which will be the limit of the present eco-label scheme (the Danish environmental protection agency (EPA) now suggest 5–30%). In any case it will be difficult for the trade associations to come to terms with a system that rewards the top performers at the expense of the rest of their membership.

It is further suggested that EMAS will yield positive environmental benefits in direct contrast to the achievement of arbitrary standards set up by the eco-label scheme which CEPI claims have been the target of political expediency.

Another reason why CEPI is not supportive is that it believes the existing national and independent schemes which are abundant among paper products will continue to operate, thus potentially diluting the benefit of an EU eco-label.

The European Tissue Symposium (ETS)

Representing tissue paper producers throughout Western Europe, the ETS has also announced its concerns about eco-labelling, particularly with regard to the lack of objectivity of the system. In its view, the recent EU decision on the eco-labelling of tissue products fails to meet the prime objectives of the original Regulation, which are stated:

- To promote the design, production, marketing and use of products which have reduced environmental impact during their life cycle.
- To provide consumers with better information on the environmental impacts of products.

The ETS points out that the eco-labelling exercise has highlighted the position that for many products such as paper, the environmental impacts are mainly production-related and thus the establishment of valid and workable criteria can be quite difficult. One of the main concerns is that the criteria as developed could be open to different interpretations by each member state. The net result is that for the same product a manufacturer could obtain approval in one country and rejection in another. One reason why this could come about is that some member states still have difficulty in developing a full life-cycle analysis (LCA). This is not surprising as the development of this complex is still in its infancy.

The ETS also believes that some government representatives have become convinced that only paper grades made from low-category waste paper should be awarded an eco-label. As a result a strong bias towards recycled products has developed without a full understanding of the environmental impact of the recycling aspect itself, including for example the increased sludge level from de-inking.

Outcome

The effect of this opposition from CEPI and ETS, whose membership spans all of the paper-producing countries in Western Europe, is not yet clear. Although one paper producer has been awarded the EU eco-label, it has not yet been generally accepted.

Even if the European producers stand firm on this issue, overseas companies may adopt a different attitude. It is reported that the Danish EPA and Director-General XI recently met with non - European paper producers, mainly from the USA, Canada and Brazil, to listen to their views. The American Forest and Paper Association (AF&PA) has expressed concern about the criteria, mainly sulphur dioxide emissions and waste, which it says are not rated as highly in the US as environmental problems. Involvement of these producers becomes even more complicated when it is remembered that a significant proportion of the European mills are now under foreign ownership.

Meanwhile, it is understood that the process within the Commission to finalise the fine papers eco-label is complete and has been formally adopted by the Commission. A third product group covering envelopes is now being evaluated.

While the outcome of all this is awaited, two key questions remain. Can the industry actually halt the eco-label system? What is the legal position for each side with regard to free trade and competition law? Now that one company has accepted eco-labelling it will be interesting to see if and when others follow.

Energy

The link between energy production and the greenhouse effect has again forced fossil fuels back into the environmental debate. As the fifth largest user of industrial energy world-wide, the paper industry will inevitably be drawn into the arguments about methods to be pursued to reduce the changes in climate through which their effects on humankind are seen.

As has already been said, the paper industry has become more deeply involved in energy management, so far to good effect, but with more work necessary. There will be more and more sophisticated CHP projects, combining efficiently the manufacture of electricity and process heat. A project was completed in 1996 for Enso, in Spain where the project cost was around US$57 million, and where the generation supplied not only the mill, but also had spare capacity which was sold to the national grid, so improving the basic economics of the mill project. It is interesting that sophisticated in-plant management and automation, at a cost of around US$1.3 million was included in the project.

Progress in other areas is under review – it is necessary to analyse the total system of energy consumption in an operation, reviewing not only those items in-plant that contribute to process efficiency, but also items like transportation through the various links. Such links cover, for example, the movement of waste paper, logs, all raw materials and all end products. Analysis will lead to discovering the way in which they may be operated so as to reduce total energy consumption. This again is driven by the requirement to reduce carbon dioxide emission. Ironically, such an analysis could militate against virgin pulp replacement by recycled fibre in some of the more remote paper mill locations.

Certainly the emphasis is on more information and more control of the operation which will be used to yield energy economies, some of which could have quite far reaching effects.

Summary

Public perception of environmental issues affect the industry. By and large the industry has taken on board the environmental concerns that have been generated, and is working to reduce the causes of concern.

Examples of legislation to control and where possible prevent pollution have been given, and the fact that expensive landfill is not the solution to problems of disposal was discussed. The advantages of the controversial EU Packaging Directive were discussed. Voluntary schemes to manage the environment are being applied in a number of countries, and the steps taken to obtain a BS 7750 certification have been dealt with.

It can be seen that there is a great deal of activity on issues concerning the environment. Legislation is tightening on water usage, and the prospect of zero effluent becomes more attractive.

It is now being suggested that for the pulp and paper industry, water conservation and energy saving will gradually overtake sustainable forestry as corporate environmental objectives.

CHAPTER

9

The challenges to the paper industry

In considering the way ahead it is worth pausing to consider the challenges to the industry, and the possible requirement to find alternatives to paper. They come from a number of quarters. First and foremost there is the environmental challenge and it is illustrated by the question 'How does the industry reconcile the use of the forests with the environmental demands in the new millennium?' The question of the viability of the product then requires discussion, particularly, 'Why is paper still continuing to survive, and will it continue to do so, or are there alternative products?' Finally the changing shape of companies and the effect of paper on the world economy need considering.

Introduction

The biggest challenge to the future of the industry is provided by the environmental challenge to the deployment of forests. The paper industry has a particular interest in all matters that affect its future, and the forest must rank high on its list of interests. This is followed by local environmental concerns, where effluent and its treatment still remains an important consideration. Much work has been carried out to increase the acceptability and the amount of recycled fibre that can be used in the product. The amount used also has an impact on both the forestry arguments and the local environmental concerns, and the technical and macro-economic aspects must both be considered. Given that these problems have been satisfactorily addressed, the future significance of paper as a raw material has been tacitly assumed and that too must be considered and questioned.

Environmental concerns

The forests

In the natural world the forests grow abundantly and little disturbs their growth or their provision of resources for other users. However, it is quite clear that a forest will only yield a high harvest if it is looked after. In the past, when forests were allowed to grow unattended they fell into some degree of dereliction and decay. In more recent years, although some forests have been well managed, many have not, and still more have been just used without thought for what will happen as a result of this use. This applies almost exclusively outside the paper industry, but is nevertheless relevant to it. The environmental groups are serving a useful purpose in assuming the role of the world's social conscience. There is a middle path through the forest that should be acceptable to all, while at the same time being beneficial to the world's economy.

All the forests of the world need to be managed. In the same way that a building needs maintenance and attention for its continued and better use, so the forest needs to be maintained and developed for its better use. The Canadian initiative to provide some assistance with this work is to be applauded. The work must be properly and expertly undertaken, and the qualified forester has an important job in carrying out and developing this work. The use of the raw material must be regulated and consideration given to those who currently use, and in some cases misuse, it. If a community relies upon the forest for fuel, it is important that provision is made for that community to continue to have an inexpensive fuel source. Perhaps plantations for wood fuel, grown in a sustainable manner like a crop, will provide an efficient answer. Then a slash and burn policy will be transformed into a harvest and regenerate policy, provided always that it is linked with sustainable forestry. This will control one significant forest use.

Equally if a group relies upon the forest for foodstuffs it is important that the reliance can continue, and that the wildlife that use the forest as a habitat can also be accommodated.

Forest management taking account of all of these factors and others needs to be more widely introduced. This of course is not an inexpensive exercise, and funds must be generated to allow good management to be put in place. The forests therefore also need to strike an economic balance, and use for reward will enable the strands of management to be put into place, since it is unlikely that the management of the forest can be carried out philanthropically. Plantations for pulpwood, properly grown and managed within the greater scheme of forest life, will also be powerful in generating good forest management. Indeed it may also generate more efficient tree growth so that the wood can be harvested more efficiently, and the demand met from a smaller area of forest.

World forests are a huge international resource. Whether or not they are privately or state owned, the fact remains that they are a valuable resource, and work carried out to ensure that more forests are well managed will result in the resource value being realised without destroying the ecosystems that are fundamental to the sustainable nature of the forest. This has been the case for many years now in some of the big forestry concerns, for example in Scandinavia, Canada and the United States, and in Australia and New Zealand. The areas where more work needs to be carried out, and carried out urgently, is in the countries that are emerging, usually from regimes that are economically poor, for example, Poland, Russia (where work is currently being undertaken) and Africa. Here whole societies are totally dependent upon the forest for wood, for warmth and for cooking, and the progress necessary will need to be introduced with sensitivity.

Suitable funding to undertake this work must be found, but the reward from such projects could be large. Not only will the financial resources be necessary, but there will be the need for more well-trained humans and in particular small teams of skilled forest workers and foresters. This is perhaps one of the greatest challenges that must be met.

The industry uses a wood-based raw material, and bad public relations connected with that raw material can only be to the detriment of the paper industry in general terms. It is not good enough to take a passive view, it is necessary to work towards an international, industry-wide view. It is possible that the paper trade could take the management of the forests under its wing, as in the Canadian initiative referred to earlier, and this would be highly desirable.

Equally, land clearance comes into a similar category of responsibility, and some industry-wide policy on this peripheral aspect of the paper business also needs consideration.

Development in the genetic engineering of trees is taking place now, and this important work will continue. In a recent article ('Paper, people and biotechnology'), Robert A. Wilson talks about the work being carried out in bioengineering. The possibilities seem again to be quite vast. Genetic engineering is currently directed at improving the growth rates of species, with all that this means in the utilisation of capital that is currently tied up for long periods in forests. Work is being undertaken on a number of species including eucalyptus and pine and has yielded examples giving up to 100% faster growth.

Besides this, some work is also directed at pulp yield and fibre morphology, at pulping and processing costs, where reductions in lignin content clearly favourably affect pulping efficiency, and also at paper properties. He suggests the possibility of creating DNA linked pulp and paper characteristics.

It could well be that the art that the papermaker has developed in the process, of modifying pulp characteristics to produce desirable paper properties will be eventually superseded by this work. This is still a distant goal, for a major new technology that was barely in existence 20 or so years ago.

Having mentioned the application of genetic engineering to forestry, it is also applicable to the agro crops - straw, hemp, kenaf. This work, coming as it does at a time when the use of agricultural land is being rethought because of the oversupply of certain crops, opens the future of raw material supply to the paper industry to change.

Other fibres

Currently around 55% of the fibre requirements of the industry are met by using virgin pulp derived mainly from the pulpwood supply. The use of straw as a twentieth century raw material for papermaking has received a considerable amount of publicity over recent years, but only a small amount, at the most 5%, of non-wood fibres such as straw and bagasse is actually used. There are still great logistics problems with straw usage although there may be a good opportunity for development of straw pulping, in particular, in rice-growing areas such as China. This is particularly pertinent where raw materials have to be imported, and are at a premium. A small mill using local straw as a raw material could develop a competitive edge, and perhaps minimills could be sited near big conurbations. Quite how this would be handled as the industry gets bigger and as the volume required increases is a matter of conjecture but this writer believes that as manufacturing concerns get larger, and the scale of the operations get bigger, the use of straw as a raw material is less likely to survive. To the dismay of the environmental lobby, the straw-pulping process is heavily polluting, which must be a negative as far as the wide-scale adoption of the raw material is concerned, However, the turning point is the unit size of the pulp and paper mill. The yield on straw pulping is of the order of 33% so that a mill with a paper output of, say, 50 000 tonnes will require a supply of straw of, say, 150 000 tonnes, all becoming available at the time of harvest, and so needing storing for subsequent use, even in countries where the straw harvest may take place a number of times in the year. The use of straw as a raw material clearly remains a challenge to the paper industry.

Other raw materials such as flax and hemp and bagasse will continue to be developed for small and often specialised operations. Genetic engineering will, in the long term, have some effect on the development of agrofibres for papermaking.

Recycling

Almost 40% of the fibre requirement is obtained from recycled materials. The use of recycled fibre will continue to expand at a faster rate than virgin fibre and by the turn of the century could well account for close to half of the fibre requirement. In several parts of the world governments are preparing legislation to make the inclusion of secondary fibre in different paper grades mandatory. For example the USA now requires the inclusion of recycled fibre in grades such as newsprint to a level of 25%, and this is expected to rise to 40%.

There will be more waste paper collected, upgraded and recycled, and this will be particularly important in some of the countries that have not so far responded seriously to the pressure for increased recycling. In part also this will be influenced by regions where big increases in the consumption of paper are anticipated. Particular areas will be Eastern Europe, where only 26% of paper production was recycled in 1993, followed by North America (33%), which has a thriving pulp-producing base.

Consequently a world trade in waste paper is developing which has already exceeded 15 million tonnes per annum. The main shipments are exports from North America to countries in the Far East and exports from Germany who are using such exports as a means of topping up the high levels of recycling to which they are committed. The Nordic region, despite its very substantial forest resources, has already started to import waste paper so as to be in a position to react to an anticipated market demand for more paper containing recycled fibre.

In some cases recycling amounts are seen to be getting quite high as in the case of newsprint, for example, where the use of recycled material approaches 100%. There more work needs to be carried out to decide whether or not there is an optimum for recycling levels. Such an optimum could come about for two distinctly different reasons.

The first would be on technical grounds. Work so far carried out has not been conclusive, but although some of it indicates quite strongly that there is no technical barrier to the number of times that a fibre can be reused, this has still to be proved. Some retain the belief that a maximum reuse of seven cycles pertains. It will be interesting to see the trends of paper properties when, for example, a newsprint mill manufacturing from 100% waste has been in operation for a number of years, and more particularly when the market in which it operates uses only newsprint made from 100% recycled waste. When this happens, some of the recycled fibre used will already have been previously recycled, possibly a number of times. There is a school of thought that the product requires the inclusion of some new fibre, and this too needs to be investigated more fully.

The product from a 100% recycled newsprint mill should be compared with that from a woodpulp-based newsprint mill over the span of a few years. This will establish whether or not the recycled product degenerates with time. One can then discover if there is in fact a threshold of virgin fibre needed to maintain the quality of the product, and this must be an area where work will continue to be carried out. However, in the last analysis not all newsprint made will be recoverable, with some sources suggesting that the maximum recovery figure possible will be only about 75%. In fact this will become even more important when the levels of recycling in the majority of other grades also approach these levels.

The second problem with increased waste utilisation is economic and concerns the effect of higher levels of recycling upon the demand for pulp and hence the resources available for forestry. It is quite clear that some of the best forest management has come about because there has been considerable investment in forest development. Investment usually occurs when there is a tangible return from that investment. The return from good forest management is the continuing health of the paper industry and the continuing availability of wood for pulp. If we are too successful in recycling, this may be counter-productive to providing resources for good forest management. This is another area where some research needs to be carried out, and may influence the way ahead over the next few decades.

Work must be carried out to provide a solution to both of these problems, so providing a view on optimum levels of recycling.

Impact of environmental issues

In addition to the effect of the pressure group campaigns, the need to comply with current and forthcoming environmental legislation will cause changes in trade patterns. For example, in Taiwan, which is now the twelfth largest consumer of paper world-wide, the cost of environmental compliance will shortly lead to the closure of the domestic pulp business which will be relocated in mainland China or Vietnam. The implication of the Endangered Species Act in the US Pacific Northwest has already resulted in very substantial reductions in the allowed forest harvest. Consequently the paper producers in that area are now looking for wood supplies elsewhere, even as far away as Russia. This move would boost the trade already developing in logs and wood chips between certain countries.

Under a new EU-IPPC Directive, the European Union – Integrated Pollution Prevention and Control, all paper mills in Europe will need a licence to continue to operate after the year 2005. This will only be granted if tight emission standards are met, mainly drawn up after consideration of the best available technology. This could lead to the siting of future paper and pulp manufacturing capacity in countries with less onerous environmental regulations.

Over-zealous application of environmental laws could well cause trade distortions, and this has certainly proved to be the case with the introduction of the German Packaging Waste Ordinance. This generated a significant surplus of waste plastic and paper, which when exported had an adverse effect upon the economics of waste collection systems being developed by neighbouring European countries, so putting into question the viability of these operations.

Utilisation of energy will continue to attract attention and work will give rise to further important reductions in the amount of energy required in the processes.

This will be accomplished not only by setting up more combined heat and power schemes, but also by the introduction of information systems detailing energy consumption throughout the complete operating system. Each mill is likely to have an employee responsible for this aspect of the business, and from this systems analysis, usage will be controlled. It is likely that many plants will have an overproduction of energy, and management of the sale and purchase of energy will therefore also be important.

To quote Simon Lloyd in his article on '*Strategy Focus on Energy Cost Saving*' in PT 37:5 '... energy efficiency with environmental benefits and cost reduction will undoubtably remain priorities for management in the paper industry as competition intensifies. ... As energy can be a complex activity, a focused approach is needed ...'.

Effluent and landfill

Landfill sites are clearly in short supply, and they are therefore expensive and, to many, undesirable. Effluent treatment is an important feature of the industry, and will continue to be so. It is imperative that work is carried out directed at researching and developing the zero effluent concept, but it should be continued in parallel with the search to find uses for the effluent itself. Incineration is a real possibility and it would of course substitute for fuel use and so provide another environmental benefit.

Landfill and alternatives

There would seem to be three ways in which the paper mill can dispose of its effluent. After settling and concentration, the sludge remaining must be disposed of. The traditional way of doing this, developed over the past 20–30 years, and in some cases more, has been to pay to have it taken away for landfill. This has become a more complex proposition with legislation demanding more knowledge about all of the materials being tipped. As already stated, landfill sites are becoming rarer, particularly in the USA where the lack of availability of sufficient sites is now quite a large problem. The result of this is that the cost of tipping is becoming hugely expensive and what had seemed to be the solution to the problem of sludge disposal is now being overtaken by subsequent events.

Two solutions are being worked on. The first is the incineration of sludge, which demands, first of all, the removal from it of as much water as possible before incineration. It is then incinerated through recovery energy boilers, and the energy so generated is reused in the process.

The second solution is more appealing and is currently being researched in a number of places. The research is aimed at developing the 'zero effluent' pulp and paper mill. It is attractive to retain as much material as possible in the paper or pulp, and indeed it is an option that has exercised the pulp and papermaker over many years. Paper machine water systems have been 'closed' for a number of years, and this has been achieved by a process of defining and isolating all water usage and replacing freshwater with recycled water. Good examples of this substitution may be found in the process where there are closed loops allowing the reuse of water to take place, for example, around pump gland sealing water or felt washing water. This careful scientific approach to freshwater replacement has yielded good cost savings. Operating on the principle that to solve 90% of the problem is the most cost-effective solution, the industry has been able, for some years now, to operate its water systems to a high degree of closure. But now the standard has changed, and it is becoming a necessity to run pulp and paper mills with further reduced effluent. There are, of course, a number of disadvantages, not least the opportunity for unwanted organisms to grow in the system, but they are likely to be outweighed by the advantages, and there will be developments to the achievement of 'zero effluent' in the next decade.

The future of paper

Why use paper? So far we have not really considered any challenge to paper, and the question needs to be addressed. Will communications backed up by the written word always take place? In this electronic age, there is the Internet and e-mail, and a feeling that the paperless office is on its way. The 'paperless office' of the 1960s has already been referred to and what finally manifested itself as far as the industry was concerned was an increase in the consumption of paper, rather than a decrease. Now with the wider acceptance of personal computers and the growth of electronic mail one could expect another bout of comment that the paperless society will now come upon us. However, people are now just beginning to talk about the effect of this on the environment, calling it the 'environmental footprint' of this computer application. It is more likely to again result in the consumption of still more paper. The fact of the matter has always been that consumers are content to have letters and books and papers, and there is no sign that this habit will change. This is particularly pertinent in the developing countries, but even in the more sophisticated countries, not everyone has a computer and very few computers are sufficiently portable to use anywhere. Can you imagine reading the electronic version of the *Financial Times* on the way to work in the morning?

Paper has a dominant position in the operation of our society. This book has been written using a PC, but this has encouraged more drafts to be printed than would otherwise have been the case. Newspapers are ever-increasing in size, although the tonnage of newsprint consumed has not risen steadily in the last decade. This was probably caused by world recession causing, for a while, reduced advertising of products in the newspapers, and by the use of lighter-weight papers. There is more junk mail in circulation, possibly again because addresses are more collectible and accessible through the computer. More affluence leads to more consumer spending, more packaging and more advertising. It is hard to imagine anything other than an increase in consumption.

Again, proper food packaging is of benefit to society, and one of the problems of bulk food distribution in some of the less developed countries is that it is not adequately packed, and so has a reduced life as a result of this.

Paper is environmentally friendly in that, unlike plastics, it is made from a renewable resource. Paper can all, or almost all, be recycled, and there may be some move back into the packaging areas that were at one time lost to plastics, which are oil-based and not readily recyclable. It is therefore a considered opinion that both the fibre supply and the paper requirement will continue to grow, and that this will continue to be the case for the foreseeable future. Within this position, the fibre supply will move in favour of the RCF component rather than virgin pulp.

Although there are signs that some per capita consumption is decreasing, or at least levelling off, for example in America, Sweden and Europe, there is much scope for increases from other countries further down the consumption scale. There is thus a very long way to travel before the industry even approaches a decline.

Industry structure

The past decade has again proved to be a difficult time for the paper industry. While paper consumption continued to rise, the period 1960 to 1975 saw the modernisation of much plant and the closure of small and inefficient operating units. This modernisation was coupled with the transition of a craft-based industry to a more technological industry, and one more able to meet the demands of the twentieth century. Then again, in the period 1980–1995, a combination of over-capacity in all of the key grades and a deep recession in almost all of the major consumer countries reduced and often removed the profitability of most of the leading companies, many of which were carrying heavy debts from the spending boom of the 1980s.

The emphasis for the capital-intensive paper industry must be to become more profitable and to get away from its cyclic nature. It must also improve its environmental profile by contributing to, and being seen to contribute to, an improved environment.

One aspect of improved profitability lies in growing the size of an operating company, usually by acquisition. The capital-intensive nature of the industry has been highlighted on a number of occasions. A modern mill, typically producing 250 000–300 000 tonnes of paper a year, will cost in the region of US$750 million to install, and if it is integrated with a pulp mill, this will take the cost to over a billion dollars. A key factor for a mill of this size must be to have secure access to a fibre supply which will hopefully be both low cost and consistent in its quality and supply.

As a result, the development of the industry to date has been strategically linked to the location of natural forests in the Northern Hemisphere, in particular North America and the Nordic region. With the majority of virgin pulp being made using wood fibre, the environmental pressure over conserving trees, especially old growth, has become a fundamental issue for the industry.

However in the 1970s, the establishment of new forests, in particular the fast-growing plantations in the Southern Hemisphere, opened up the possibility of a new source of wood supply based on eucalyptus trees, from regions such as South America and Iberia. Following this lead, a significant part of the next wave of pulp capacity is likely to be installed in the Pacific Rim countries such as Indonesia, where conditions are also favourable for fast tree growth. This will inevitably lead to forest, pulping and papermaking integration led by the existing major operators in the business. Neither must the urban forest be ignored and the manufacture of many commodity grades, such as newsprint and corrugating medium, will be expanded near their principal raw material collection areas, which will be close to the major conurbations.

Consumption

Despite previous predictions of the possibility of the paperless society, paper has maintained steady growth for many decades, easily keeping up with the growth of gross domestic product (GDP). The USA accounts for just over 30% of the world pulp and paper market both in terms of production and consumption. It is therefore no surprise that one of the most important growth indicators in the world of paper consumption has historically been the health of the US economy.

In the future, further threats will come from the development of information superhighways and the environmental pressure to reduce overall consumption and save resources. However, any slow down in consumption in the developed world is likely to be easily matched by growth from the emerging countries, particularly from those in Asia. The Asian share of total world consumption has been rising at a rate of about 1% per year for several years, and this reflects the very fast growth in economic development especially in South East Asia. If present growth rates are maintained, China will become the second largest world-wide consumer of paper after the USA soon after the turn of the century.

Just as the first chapter contained a statement about the size of the industry, so it is appropriate to make some comment on its future growth. Even if the industry growth is only at an average 3% per annum, consumption will double every 24 years or so, with 300 million tonnes of paper consumed expected to be passed by the year 2000.

There is no paper substitute recognised at the present time. Figure 9.1 shows three world consumption growth scenarios. There is an 'actual' line, and this fits very closely the line of 4% growth calculated from an actual datum point in 1975. There is also the 3% growth line calculated from the same datum. The figure lends much credibility to the forecast, that the industry can expect overall an average growth to continue at around 3% per annum, although a more optimistic projection of 3.5% to 4% would not be unreasonable. The growth varies from year to year. For example in 1994–95 growth was 3.3%.

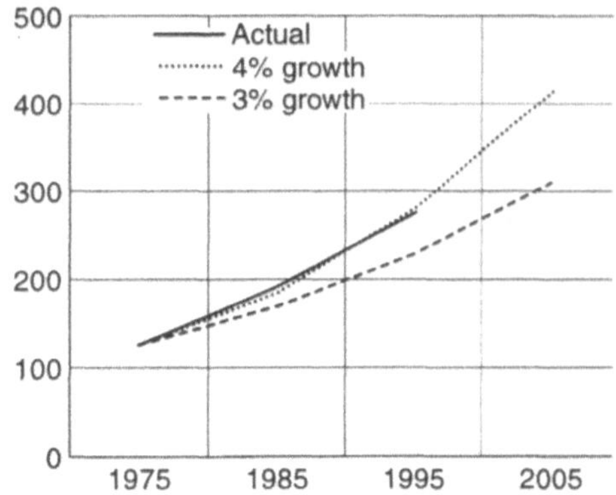

9.1 Calculated growth in consumption with growth at 3% and 4% per annum v. actual consumption.

World economy

Company economy has been dealt with, but once again the contribution of the paper trade to the world economy is a factor of some importance. A crude rule of thumb figure was developed in Chapter 1, and it is worth reminding ourselves of it. The figure derived for world trade was US$500 billion, making the paper industry an important contributor to the world economy. As stated above, the increase of consumption is expected to rise at a conservative 3% per annum. This will make an important contribution to the growth of the total world economy.

It is an irony that profitability is often interfered with, as when for example a big producing machine coming on stream, flooding the market, and upsetting the price structure of that market for a considerable time. The business must also be able to control the way in which the market is fed by the capacity from such new capital expenditure. It is here that the strategy of company acquisition is important.

The industry must discover how to manage the advent of such a large tonnage of paper better so that it does not adversely affect profitability. This is another reason for the growth of companies by acquisition abroad.

Public relations

It is of prime importance that the industry must first be in a position where it cannot be criticised, particularly in the environmental field, and it must then tell the public of its record. The tissue companies have shown the way to achieving this goal, as have some of the companies involved in forestry operations. For too long has the industry tried to maintain a low profile in the public perception. It has a good story to tell and it should tell it. For many years it was so concerned that the public would associate the use of waste paper with a cheap product that it preferred not to tell its story of recycling, and that allowed the environmentalists the opportunity to criticise the industry for consuming raw material without concern about the environment, and without having the statement challenged by the industry. This prejudice has at last been overcome.

Summary

The challenge to the industry that is provided by environmental concern, particularly for the forests, highlights the need for more work to improve some forest management. Plantation growth and genetic tree engineering will have a place in this.

Straw continues to attract attention as an alternative raw material for paper manufacture, but it has some serious disadvantages. Increasing the amount of recycled fibres remains important, and alternatives to effluent disposal by landfill are sought, with incineration and zero effluent appealing.

Paper will continue to be a growing industry, but steady profitability needs to be addressed.

Finally, good public relations are an imperative.

CHAPTER

10

Summary of main points

A brief summary of the main points developed in the book is given. It uses the chapter headings or subheadings as a framework, to present the essential points.

Overview of the paper industry

The book has set out to give a broad appreciation of the international paper trade. It also seeks to give an indication of what are the main drivers that cause change to be made, and how the paper industry is changing. In doing this a view has been taken of the development of the trade and of the ways in which the industry may change in the next decade. This chapter is intended to focus briefly upon the main points that have been discussed. The views set out are based on a broad understanding of the industry and the current drivers of change.

The relevant chapter headings or subheadings now form the structure of this chapter, but the conclusions are not intended as a plan for research and development for the industry.

Table 10.1 Global value of paper components

Paper	(275 million tonnes)	US$350 billion
Pulp	(175 million tonnes)	US$120 billion
Recycled fibre	(100 million tonnes)	US$15 billion
Wood for pulp	(650 million m^3)	US$30 billion
Minerals	(20 million tonnes)	US$4 billion
Chemicals		US$15 billion

Scope of the industry

The world paper trade comprises trade in pulp and paper both within a country and between countries. A multi-billion dollar trade has been described. In 1995 world consumption was 276 million tonnes, and almost a half of this - actually 46% - was produced in North America and Europe.

Paper is made up of the main components virgin pulp (57%), RCF (36%) and mineral (7%), and water is of great importance to its manufacture. The values of these items of the trade are crudely estimated in Table 10.1. The total direct value is of the order of US$534 billion and that does not take account of the indirect values contributed by the principal suppliers to the trade, notable among them being the suppliers of chemicals, machinery and machine clothing.

As much as 20% paper production enters the world import and export trade, generating business worth around US$80 million. The paper industry accounts for 2.5% of the world's production and 2% of world trade. Besides this it has a considerable social significance through its contribution to education and health.

The business is a truly international trade and it provides direct employment for upwards of a million people.

The long-term growth of paper consumption is expected to continue at around 3% per annum, and the volume of paper consumed world-wide will exceed 300 million tonnes by the turn of the century, or very soon after. The growth rate is at a figure of 9 million tonnes per annum.

Process changes

The industry is highly capital-intensive, and the manufacturing process is unlikely to change radically in the next decade. Pulp mills will continue to develop chlorine-free bleaching methods, driven by environmental reasons. Paper machines will edge up in speed and width.

A 'state of the art' paper machine is broadly 10 metres wide and 350 metres long. It produces close to 300 000 tonnes per annum, and costs on a green-field site around US$750 million. If the machine is integrated with a pulp mill, the cost could rise to over a billion US dollars.

Consumption

World population is increasing fastest in Africa, Latin America and Asia. The per capita consumption of paper varies from 2.7 kg per person per annum in Africa to 333 kg per person per annum in the USA, with a world average of 48.6 kg per person per annum.

Asia, and particularly China and Indonesia, is the most important paper consumption growth area, followed by Latin America. Eastern Europe and Africa will grow, but not vigorously, until their economic climate has improved.

Paper grades showing the fastest growth are corrugating materials, printings and writings, and tissue. Newsprint will show significant growth, particularly in Asia.

Forestry

Some world forests are well managed, but the management of some leave much to be desired. They are scarcely managed at all and have become derelict. Some well-managed forests in the industry are well on the way to being environmentally sound and sustainable. Attention must be paid to the forests that are not well managed, even where they do not actually provide raw materials for paper. A lead to this approach is given by Canada, where the forests are of great importance to the country's economy.

North America provides almost 50% of the world's pulp supply.

Although they are not problems directly influencing the provision of pulp wood, the question of forests as providers of fuel must be addressed, as must the clearing of forest for agricultural purposes. In all applications, plantations give far better timber yields than natural forests.

The next decade will see an advance in the genetic engineering of trees for pulping, and this work is all-important for the development of the pulp and paper industry.

Raw materials

The provision of straw and other similar fibres as a papermaking raw material will continue to be explored. Biotechnology will be developed and the use of enzymes and similar materials will assist the development of genetic engineering in the forests. This will lead to some substantial improvements, in forestry, pulping, papermaking, de-inking and disposal of effluent. The systematic reduction of energy consumption will continue.

Recycling

The consumption of waste paper will continue to increase, driven in some places by legislation. It is a real possibility that the amount of recycled fibre, currently at 38% of the fibre supply, will reach 50% in the next ten year period.

More work is required to explore again the technical limits of recycling, and the economic question raised by the possible adverse effect that increasing percentages may have in detracting from work on forest management.

The international trade in waste paper will continue to grow, and the provision of more production capacity will continue to be a target for investment capital. The urban forest will have its effect upon the location and installation of new plant.

Sludge disposal by incineration will continue to attract more industry attention.

Environmental issues

Environmental issues will continue to drive the direction of development of the industry. Public relations will be of increasing importance and the industry must develop its ways of getting its achievements across. The industry uses a naturally renewable raw material. It can be effluent-free, and the product is sustainable and biodegradable.

Water and effluent treatment will continue to consume investment capital. Work will continue to explore the possibility of zero effluent mills.

Timber certification will continue to be an issue, as will directives aimed at controlling the environment through legislative measures. Voluntary schemes are being developed, for example BS 7750 and ISO 14000, and these will have their place in making the industry environmentally sound.

Taking all factors into account, the paper trade is strong and robust, and should continue to retain a viable place in the world, and be a strong contributor to the economy of the world.

APPENDIX

I

World paper industry trade federations

Extracted from Birkner Paper World 1996-7, a directory of the International Paper World, by kind permission of the publisher.

Argentina
Asociacion de Fabricantes de Celulosa y Papel,
Avenida Belgrano 2852,
1209 Buenos Aires
Fax: 54 1 97 00 53

Austria
Austro papier Vereinigung
Osterreichisalier Papier industrieller,
Gumpeldorfer Strasse 6, A 1061
Vienna
Fax: 43 222 58886 222

Australia
Pulp and Paper Manufacturers
Federation of Australia,
PPMF,
Royal Domain Centre, 368 St Kilda
Road, Melbourne, Victoria 3004
Fax: 3 690 11 80

Belgium
Association des Fabricantes de Pates,
Papiers et Cartons de Belgique,
COBELPA,
Avenue Louise 306, B-1050 Bruxelles 5,
Fax: 2 646 82 97

Brazil
Associacoa National dos Fabricantes
de Papel e Celulose, ANFPC, Rua
Afonso de Freitas, 499 Paraiso 04006-
900,
Sao Paulo SP.
Fax: 11 88 53 689

Canada
Canadian Pulp and Paper Association,
CPPA,
Sun Life Building, 19th Floor,
1155 Metcalfe Street, Monteal,
Quebec H3B 4T6
Fax: 514 866 3035

China
China Paper Association, CPA,
Huang Si Da Jie Jia 3, Beijing
Tel 10 6752 104

Czech/Slovak Republic
Svaz Prumyslu papiru a celulozy,
Czech Pulp and Paper Association,
Skretova 6, CZ-12059 Praha 2.
Fax 2 242 13208

Denmark
Sammenslutuingen af Danske Papir,
Pap-og Cellulose fabrikker,
Danish Paper Board and Pulp Makers Association,
Dannisco Paper A/S, Korsgade 22,
DK-8500 Grena
Fax: 86 32 54 77

Finland
Finnish Forest Industries Federation,
Etel desplanadi 2, FIN-00130,
Helsinki 13
Fax: 09 17 44 79

France
UTIPATES
Association Francaise des Utilisateur de Pates,
154 Boulevard Haussman,
F-75008, Paris
Fax: 1 45 62 82 47

Germany
Arbeitgeberverband der Deutschen Tapetenindustrie,
Lindenring 44, D-60431, Frankfurt
Fax: 69 52 00 36

Great Britain
The Paper Federation of Great Britain,
Papermakers House, Rivenhall Road, Westlea,
Swindon SN5 7BE
Fax: 44 1793 886 182

Hungary
Papir-es Nyandaipari Muszaki, Egzesulet,
Fo utca 68, Budapest H-10278
Fax: 1 201 02 56

India
Indian Paper Mills Association
India Exchange, 4 India Exchange Place,
Calcutta 700001
Fax: 33 223 242

Indonesia
Indonesian Pulp and Paper Association
Jalan Cimandiri 6, Flat 1 /2,
Jakarta 10330
Fax: 21 320 168

Italy
Associazione Italiana fra gli Industriali della Carta, Cartoni e Paste per Carta,
ASSOCARTA,
Bastioni de Porta, Volta 7,
1-20121, Milano
Fax: 2 2900 3396

Japan
Japan Paper Association
9-11 Ginza 3-chome, Chino-Ku,
Tokyo 104
Fax: 3 3249-4826

Korea
Korea Paper Manufacturers Association
Room 302 Songpa Building, 505 Shinsa-Dong, Kangnam-gu,
Seoul 135-120
Fax: 2 549 0980

Malaysia
Malaysia Pulp and Paper Manufacturers Association
Suite 710, 7th Floor, Wisma Lim Foo Yong,
86 Jalan Raja Chulan, 50200 Kuala Lumpur
Fax: 3 775 1519

Netherlands
Vereniging van Nederlandse Papier en Carton fabtrieken
Julianastraat 30, 2012 ES, Haarlem
Fax: 23 532 68 98

New Zealand
New Zealand Pulp and paper Industry Association Inc.,
Trust Bank Building, 7th Floor,
Hinemoa St., Rotorua
Fax: 7 347 1376

Norway
Skogsbrukets og Skogindustriens Forskningsforening,
SSFR
Postpoks 24, Blindern N-0313, Oslo
Fax: 64 94 29 80

Poland
Stowarzyszenie Papiernikow Polskich,
SPP,
Plac Komung Parzskiej 5a, 90-950, Lodz
Fax: 42 32 43 65

Portugal
Associacao da Industria Papeleira,
CELPA,
Rua Marques de Sa da Bandeira 74-1 Esq.,
P-1050 Lisbon
Fax: 1 795 94 56

South Africa
TAPPSA Technical Association Pulp and Paper South Africa,
PO Box 10606,
Marine Parade
Durban, 4056 Natal

Spain
Asociacion Nacional de Fabricantes de Pastas Papeleras,
ASPAPEL,
Alcala 85-4 Piso, E 28009 Madrid
Fax: 1 577 47 10

Sweden
Skogs Industrierna,
Swedish Forest Industries Association,
Storgatan 19, S-11451 Stockholm
Fax: 8 661 73 06

Switzerland
Arbeitgeberverband Schweizerischer Papier Industrieller,
ASPI'
Bergstrasse 110,
CH-8032 Zurich
Fax: 1 252 38 82

Thailand
Thai Pulp and paper Industries Association
2nd Floor, Building 4, 1 Siam Cement Road,
Bangsue, Bangkok 10860
Fax: 2 587 0738

United States of America
Technical Association of the Pulp and Paper Industry
TAPPI
15 Technology Park, PO Box 105113,
Atlanta, Georgia 30338
Fax: 404 446 6947

Venezuela
Asociacion Venezolana de Productores de Pulpa, Papel y Carton,
Apropaca,
Edificio Johnson and Johnson, Piso 2,
Oficina 2B,
Avenida Romulo, Gallegos, Los Dos Caminos,
Caracas
Fax: 2 234 6542.

APPENDIX

II

World paper industry technical associations

Argentina
Chief Staff Officer, Arturo M Freyre,
ATIPCA,
Asociacion de Tecnicos de la
Industria Papelera y
Celulosica Argentina, Avda Belgrano
2850, Buenos Aires-CP 1209
Fax: 54 1 97 9916

Austria
General Secretary, C Zettl,
OZEPA,
Osterreichische Vereinigung der
Zellstoff-und
Papierchemiker und Teckniker,
(The Austrian Association of Pulp and
Paper Chemists and Technicians),
Gumpeldorfer Strasse 6, A-1061
Vienna
Fax: 43 222 58886 222

Australia
Executive Director, Peter J Brown,
APPITA,
The Technical Association of the
Australian and New Zealand Pulp and
Paper Industry,
Suite 47, Carlton Clocktower,
247–255 Drummond Street,
Carlton, Victoria 3053
Fax: 3 93 48 12 06

Belgium
General Secretary, A Rose,
COBELPA,
Association des Fabricants de Pates,
Papiers et Cartons de Belgique,
(Association of the Belgium Pulp,
Paper and Board Manufacturers),
39–41 Rue D'Arlon, B-1040, Brussels

Brazil
Chief Staff Officer, Sheik Rashid,
ABTCP,
Associacao Brasileira Tecnica de
Celulose e Papel,
Rua Ximbo, 165 Aclimacao BR 04108
Sao Paulo

Canada
Chief Staff Officer, Rob Wood,
CPPA-TS

Canadian Pulp and Paper Association Technical Section,
Sun Life Building, 19th Floor, 1155 Metcalfe Street,
Montreal, Quebec H3B 4T6
Fax: 1 514 866 3035.

Chile
General Secretary, Miguel L Mumos,
ATCP,
Asociacion Tecnicos de la Celulosa y el Paper,
(Technical Pulp and Paper Association),
Barros Arana, 39–2 piso, Casilla 74 C, Concepcion

China
Chief Staff Officer, X Zhong,
CTAPI,
China Technical Association of the Paper Industry,
12 Guanghua Road, Beijing
Fax: 86 6500 6577

Columbia
Executive Director, Dr Orlando A Giraldo,
Asociacion Columbiana de Tecnicos de las Industrias de Pilpa, Papel y Carton Papelsa,
Apartado Aereo No 67118, Medellin
Fax: 574 411 4967

Czech/Slovak Republics
General Secretary, Z Janda,
SPPC,
Spolechost Prumyslu Papiru a Celulozy,
(Czech Paper Technical Association),
K HRUSOVU 4, 10233 Prague 10
Fax: 42 2 707 1135

Finland
General Secretary, E Kihlman,
Suomen Paperi-Insinoorien Yhdistys (SPY) Finska Pappersingeniorsforeningen.
(The Finnish Paper Engineers Association),
Unioninkatu 17, SF 00130 Helsinki
Fax: 358 063 03 65

France
General Secretary, D Gomez,
Association Technique de l'Industrie Papetiere (ATIP),
(Technical Association of the French Paper Industry),
154 Boulevard Haussman, F-75008 Paris
Fax: 33 1 45 63 53 09

Germany
General Secretary, R Weidenmuller,
Verein der Zellstoff-und Papier-Chemiker und-Ingenieure eV (ZELLCHEMING),
(Association of Chemists and Engineers of the Pulp and Paper Industry),
Berlin Allee 56, D-6100 Darmstadt
Fax: 49 6151 31 10 76

Great Britain
Executive Director, John Clewley,
Paper Industry Technical Association (PITA),
5 Frecheville Court, Bury, Lancashire BL9 0UF
Fax: 44 161 764 5353

Hungary
General Secretary, G Madai,
Papir-es Nyomdaipari Musaki Egyesulet (PNYME)
(Technical Association of the Paper and Printing Industry),
Fo. u. 68, H 1027 Budapest
PO Box 433 H 1371 Budapest
Fax: 36 1154 090

India
Secretary, N K Naithani,
IPPTA Indian Pulp and Paper Technical Association,
Institute of Paper Technology Campus,
Saharanpur-247001

Italy
Associazione Tecnica Italiana per Celulosa e la Carta (ATICELCA),
(The Technical Association of the Italian Pulp and Paper Industry),
Via Sandro Botticeli no 19, I-20133 Milano
Fax: 39 2 266 41 41.

Japan
Executive Director, K Iida,
Japanese Technical Association of the Pulp and Paper Industry (Japan TAPPI),
c/o Kami-Parupu Kaikan Building,
9–11, Ginza, 3-Chome, Chuo-Ku,
Tokyo 104
Fax: 81 3 541 8354

Korea
Chief Staff Officer, Hak-Lae Lee,
Korean Technical Association of the Pulp and Paper Industry, Dept of Forest Products,
College of Agriculture and Life Sciences,
Seoul National University, Suwon, 441-744,
South Korea

Mexico
General Secretary, L Richardo Macias,
ATCP Asociacion Mexicana de Tecnicos de las Industrias de la Celulosa y del Papel,
AC Lafayette 138, Anzurez, 11590 Mexico DF
Fax: 52 5 203 8521

Netherlands
Vereniging van Nederlandse Papier- en Kartonfabrieken
Research and Technologiegroep,
(Association of Dutch Paper and Board Manufacturers Research and Technology Group),
Julianastraat 30, NL2012 ES Haarlem

Norway
General Secretary, K Johnsen,
Papirindustriens Tekniske Forening (PTF),
(The Technical Association of the Norwegian Pulp and Paper Industry),
Forskningsveien 3, N 0371 Oslo 3,
Box 115, Vinderen, N 0319, Oslo 3
Fax: 47 2 46 80 14

Poland
General Secretary, Z Fornalski,
Stowarzyszenie Inzynierow i Technikow Przemyslu Papierniczego w Polsce (SITPP),
(Technical Association of the Polish Paper Industry),
Plac Komumy Paryskiej 5 a,
PO Box 200,
PL 90 950 Lodz

Portugal
General Secretary
Associacao Portuguesa dos Tecnicos das Industrias de Celulose e Papel (TECNICELPA),
(The Portuguese Association of Pulp and Paper Technicians),
Rua Infantaria 15, No 9, 2 andar,
p-2300 Tomar
or
Av Dr Mario Montikho, u°20 Encosta do Restelo,
P1400, Lisboa

Slovenia
Drustvo Inzenirjev in Tehnikov Papirnistva (DITP),
(Technical Association of the Paper Industry),
YU-61000 Ljubljana, Bogisiceva 8

South Africa
General Secretary, J J Casey,
TAPPSA. Technical Assocation Pulp and Paper South Africa,
PO Box 10606, Marine Parade,
Durban, 4056 Natal

Spain
General Secretary, J L Asenjo Martinez,
Asociacion de Investigacion Tecnica de la Industria Papelera Espanola (IPE),
(The Technical Association of the Spanish Paper Industry),
Carr de la Coruna, km 7, E 28040 Madrid
PO Box 33045, E 28040 Madrid
Fax: 34 1 357 28 28

Sweden
General Secretary, B Jakobsson,
Svenska Pappers-och Cellulosaingeniorsforeningen (SPCI),
(The Swedish Association of Pulp and Paper Engineers),
Villagatan 1, S11432 Stockholm
Fax: 46 8 11 54 19

United States of America
Executive Director, W Gross,
TAPPI
Technical Association of the Pulp and Paper Industry,
15 Technology Park, PO Box 105113, Atlanta,
Georgia 30338
Fax: 404 446 6947

Venezuela
Asociacion Venezolana de Tecnicos en Celulosa y Papel Edif,
Camara de Industriales, 6 Piso, Esq Puente Aranco,
Caracas

APPENDIX

III

Commonly used units and abbreviations

AF&PA	American Forest and Paper Association
AKD	Alkyl ketene dimers
AOX	absorbable halogen
BAT	best available techniques
BOD	biochemical oxygen demand
BS	British Standard
CBI	Confederation of British Industry
CMC	carboxy methyl cellulose
Cefic	European Chemical Association
CEN	Comité Européen Normalisation, the European standardisation body
CEPI	Confederation of European Paper Industries
CHP	Combined heat and power
CIS	Commonwealth of Independent States
CMP	Chemimechanical pulp
COD	Chemical oxygen demand
CTMP	Chemithermal mechanical pulp
DSD	Dual System Germany (Deutsche)
ELV	emission limit values
EMAS	European Union Management Standard
EMS	Environmental Management scheme
EOX	Solvent-extractable organic chlorine
EPA	environmental protection agency
EQS	environmental quality standards
ESA	Endangered Species Act
ETS	European Tissue Symposium

EU	European Union
Eucepa	European Liaison Committee for Pulp and Paper
FAO	United Nations Food and Agriculture Organisation
FSC	Forest Stewardship Council
FUSSR	Former Union of States of Soviet Russia
GDP	gross domestic product
g/m^2	weight in grams of a square metre of paper
grammage	weight in grams of a square metre of paper
HMSO	Her Majesty's Stationery Office
IIASA	International Institute for Applied Systems Analysis
IIED	International Institute for Environment and Development
IPC	integrated pollution control
IPPC	integrated pollution prevention and control
ISO	International Organization for Standardization
KLS	kraft-lined strawboard
km^2	square kilometres
lbs	pounds avoirdupois
LCA	life-cycle analysis
m^2	square metres
m^3	cubic metres
min	minute
MJ	megajoules
MSW	municipal solid waste
OBAs	optical bleaching agents
PITA	Paper Industry Technical Association
PVA	polyvinyl alcohol
ppm	parts per million
PPI	Pulp and Paper International
RCF	recycled fibre
RMP	Refiner mechanical pulp
RR	renewable resources
SALP	Self-adhesive label paper
SGWP	Stone groundwood pulp
TAPPI	Technical Association of the Pulp and Paper Industry (USA)
TMP	Thermomechanical pulp
TOCl	Total organic chlorine
tonnes	2200 pounds or 1000 kg
TOX	Total organic halogen
WWF	World Wide Fund for Nature
μm	micrometres

Bibliography

Atchison Joseph E, 'Twenty-five years of global progress in nonwood plant fibre repulping', *TAPPI Journal* **79**(10) 1996.

Anon, 'Canada - an eco friendly supplier of market pulp', *Paper Technology*, Environmental Supplement, **36**(4) 1995.

Aslund A, 'Prospective markets for forest products in Eastern Europe and Russia', *Paper Technology*, **34**(4), 39 1993.

Barber M (1992), 'Latest developments affecting machine design', PITA Annual Conf.

Capps C, private communication, 1996.

Clarke D, (1995): 'Trends in paper recycling in Europe' Recycling '95 Eucepa.

Cutler P, 'Paper consumption patterns and growth', *Paper Technology*, **36**(9), 36 1995.

Friends of the Earth, 'Rainforest', 1995.

DTI: 'Biotechnology means business', Status Report, BMB Initiative, Biotechnology Directorate, London, 1997.

Gottsching L (1995), 'The German scene and the influence of legislation', Recycling '95, Eucepa.

Harriman et al, 'Bleaching of woodpulp in the 1990s in Western Europe', Harriman Chemsult Ltd, Price and Pearce Group, The Gascoigne Consultancy.

IIED (1996): 'Towards a sustainable paper cycle'.

Inman W D, 'An update on environmental affairs' *Paper Technology*, **37**(4), 2 1996.

Inman W D, 'An environmental standard' *Paper Technology*, **33**(11), 16 1992.

Jordan H, 'Environmental regulations, pollution control and energy consumption in pulp and paper', *Paper Technology*, **36**(4), 3 1995.

Kaye S, (1991), 'An energy overview' PITA Annual Conference '91.

Kirkpatrick N, (1995): 'Future environmental issues', Recycling '95, Eucepa.

Lloyd S, 'Strategy focus on energy cost saving' *Paper Technology* **37**(5), 29 1996.

Moran J, (1995) 'North American recycling', Recycling '95 Eucepa.

Nilsson S (1996) 'Review: Do we have enough forest?' in FAO Yearbook *Sources of Pulpwood.*

Saarenmaa L, 'The ecology of boreal forests and management practices in Scandinavia', *Paper Technology*, **36**(4), 34 1995.

PP&I, *PPI International Fact and Price Book 1996*, Miller Freeman.

UNCTAD, *United Nations Conference on Trade and Development Commodity Yearbook 1995.*

Waight D G, 'Health and safety - shaping of future regulations', *Paper Technology*, **32**(6), 19 1991.

Wilson R A and Martin C G: 'The paper tree', *Paper Technology*, **32**(2), 26 1991.

Wilson R A, 'Fibre supplies - a competitive advantage', *Paper Technology*, **37**(9), 23.

Wilson R A, 'Paper, people and biotechnology', *Paper Technology* **37**(3) 1996.

Index